D0102107

Inspection & Testing

GUIDANCE NOTE

621.3

DC028854

3

IEE Wiring Regulations
BS 7671 : 2001 Requirements for Electrical Installations
Including Amd No 1 : 2002

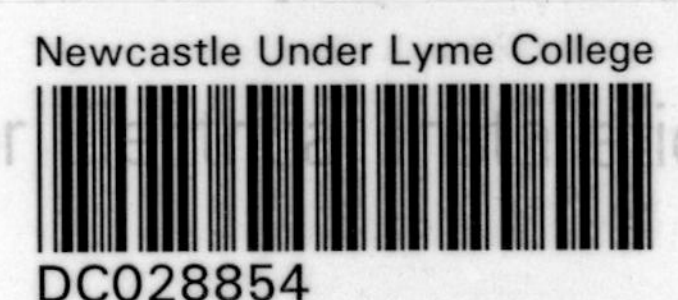
Newcastle Under Lyme College
DC028854

Published by: The Institution of Electrical Engineers, Savoy Place, LONDON, United Kingdom. WC2R 0BL

©2002: The Institution of Electrical Engineers, London

Issued August 1992
Reprinted April 1993, with amendments (Section 17)
2nd edition incorporating Amendment No 1 to BS 7671 June 1995
3rd edition incorporating Amendment No 2 to BS 7671 1997
4th edition incorporating BS 7671 : 2001 inc Amd No 1
Reprinted August 2003, new cover only

Copies may be obtained from:

The IEE
PO Box 96, STEVENAGE,
United Kingdom. SG1 2SD

Tel: +44 (0)1438 767 328
Fax: +44 (0)1438 742 792
Email: sales@iee.org
http://www.iee.org/Publish/Books/WireAssoc/

All rights reserved. No part of this publication may be reproduced, stored in a retrieval system or transmitted in any form or by any means — electronic, mechanical, photocopying, recording or otherwise — without the prior written permission of the publisher.

While the author, publisher and contributors believe that the information and guidance given in this work is correct, all parties must rely upon their own skill and judgement when making use of it. Neither the author, the publisher nor any contributor assume any liability to anyone for any loss or damage caused by any error or omission in the work, whether such error or omission is the result of negligence or any other cause. Where reference is made to legislation it is not to be considered as legal advice. Any and all such liability is disclaimed.

References to, and forms from, BS 7671 are used with the joint kind permission of the BSI and IEE

ISBN 0 85296 991 0, 2002

Contents

Co-operating Organisations

The Institution of Electrical Engineers acknowledges the contribution made by the following representatives of organisations in the preparation of this Guidance Note.

British Electrotechnical and Allied Manufacturers Association	R Lewington
British Standards Institution	W E Fancourt
City and Guilds of London Institute	H R Lovegrove
Department of the Environment, Transport and the Regions	T King
Electrical Contractors' Association	D Locke
Electrical Contractors' Association of Scotland (SELECT)	D Millar
Electricity Association	D Start
GAMBICA Association Ltd	K Morriss
Institution of Electrical Engineers	W R Allan (Editor) P R L Cook P E Donnachie D W M Latimer B J Lewis W H Wright
National Inspection Council for Electrical Installation Contracting	J Ware
Royal Institute of British Architects	J Reed
Safety Assessment Federation Ltd	N King

Preface and Scope

This Guidance Note is part of a series issued by the Wiring Regulations Policy Committee of the Institution of Electrical Engineers to simplify some of the requirements of BS 7671 : 2001 inc Amd No 1, Requirements for Electrical Installations (IEE Wiring Regulations Sixteenth Edition). Significant changes made in this 4th edition of the Guidance Note are sidelined. Note this Guidance Note does not ensure compliance with BS 7671. It is a simple guide to some of the requirements of BS 7671 but electricians should always consult BS 7671 to satisfy themselves of compliance.

The scope generally follows that of Part 7 of BS 7671 that is the inspection and testing of electrical installations in buildings. The inspection and testing of electrical equipment such as appliances is outside the scope of BS 7671 and this guide. Guidance on this is given in IEE publication "Code of Practice for In-Service Inspection and Testing of Electrical Equipment".

The principal section numbers of BS 7671 are shown on the left. The relevant Regulations and Appendices are noted in the right-hand margin. Some Guidance Notes also contain material not included in BS 7671 Requirements for Electrical Installations but which was included in earlier editions. All Guidance Notes contain references to other relevant sources of information.

Electrical installations in the United Kingdom which comply with BS 7671 are likely to achieve conformity with Statutory Regulations such as the Electricity at Work Regulations 1989, but this cannot be guaranteed. It is stressed that it is essential to establish which Statutory and other Regulations apply and to install accordingly. For example, an installation in premises subject to licensing may have requirements different from, or additional, to BS 7671, and those requirements will take precedence.

Introduction

This Guidance Note is principally concerned with Part 7 of BS 7671 — Inspection and Testing.

Neither BS 7671 nor the Guidance Notes are design guides. It is essential to prepare a full specification prior to commencement or alteration of an electrical installation. The specification should set out the detailed design and provide sufficient information to enable competent persons to carry out the installation and to commission it. The specification must include a description of how the system is to operate and all the design and operational parameters. It must provide for all the commissioning procedures that will be required and for the provision of adequate information to the user. This will be 514-09
by means of an operational manual or schedule.

It must be noted that it is a matter of contract as to which person or organisation is responsible for the production of the parts of the design, specification and any operational information.

The persons or organisations who may be concerned in the preparation of the specification include:

The Designer
The Installer
The Supplier of Electricity
The Installation Owner and/or User
The Architect
The Fire Prevention Officer
Any Regulatory Authority
Any Licensing Authority
The Health and Safety Executive

In producing the specification advice should be sought from the installation owner and/or user as to the intended use. Often, as in a speculative building, the intended use is unknown. The specification and/or the operational manual must set out the basis of use for which the installation is suitable.

Precise details of each item of equipment should be obtained from 110-03-01
the manufacturer and/or supplier and compliance with appropriate 511
standards confirmed.

The operational manual must include a description of how the system as installed is to operate and all commissioning records. The manual should also include manufacturers' technical data for all items of switchgear, luminaires, accessories, etc and any special instructions that may be needed. The Health and Safety at Work etc Act 1974 Section 6 and the Construction (Design and Management) Regulations 1994 are concerned with the provision of information, and guidance on the preparation of technical manuals is given in BS 4884 (Specification for technical manuals) and BS 4940 (Recommendations for the presentation of technical information about products and services in the construction industry). The size and complexity of the installation will dictate the nature and extent of the manual.

Section 1 — General Requirements

1.1 Safety

711 Electrical testing inherently involves some degree of hazard. It is the inspector's duty to ensure his own safety, and that of others, in the performance of his test procedures. The safety procedures detailed in Health and Safety Executive Guidance Note GS38 (revised) 'Electrical test equipment for use by electricians' should be observed.

When using test instruments, safety is best achieved by precautions such as:

(i) understanding the equipment to be used and its rating

(ii) checking that all safety procedures are followed

(iii) checking that the instruments being used conform to the appropriate British Standard safety specifications. These are BS EN 61010 Safety requirements for electrical equipment for measurement, control, and laboratory use and BS 5458 : 1978 (1993) Specification for safety requirements for indicating and recording electrical measuring instruments and their accessories. BS 5458 has been withdrawn, but is the standard to which older instruments should have been manufactured

(iv) checking that test leads including any prods or clips used are in good order, are clean and have no cracked or broken insulation. Where appropriate, the requirements of the Health and Safety Executive Guidance Note GS38 should be observed for test leads. This recommends the use of fused test leads aimed primarily at reducing the risks associated with arcing under fault conditions.

Particular attention should be paid to the safety aspects associated with any tests performed with instruments capable of generating a test voltage greater than 50 V, or which use the supply voltage for the purposes of the test in earth loop testing and residual current device (RCD) testing. Note the warnings given in Section 2.7.14 and Section 4 of this Guidance Note.

Electric shock hazards can arise from, for example, capacitive loads such as cables charged in the process of an insulation test, or voltages on the earthed metalwork whilst conducting a loop test or RCD test. The test limits quoted in these guidelines are intended to minimise the chances of receiving an electric shock during tests.

1.2 Required competence

712 The inspector carrying out the inspection and testing of any electrical installation must, as appropriate to his or her function, have a sound knowledge and experience relevant to the nature of the installation being inspected and tested, and to the technical standards. The inspector must also be fully versed in the inspection and testing procedures and employ suitable testing equipment during the inspection and testing process. 711-01-01 712-01-02

741 It is the responsibility of the inspector: 741-01-04

1) to ensure no danger occurs to any person, livestock or damage to property
2) to compare the inspection and testing results with the design criteria
3) to take a view on the condition of the installation and advise on remedial works
4) in the event of a dangerous situation, to make an immediate recommendation to the client to isolate the defective part.

1.3 The client

1.3.1 Certificates and Reports

741 Following initial verification of a new installation or changes to an 741-01-01
742 existing installation, an Electrical Installation Certificate, together 742-01-01
743 with a schedule of inspections and a schedule of test results, is 743-01-01
744 required to be given to the person ordering the work. In this context 744-01-01
"work" means the installation work not the work of carrying out the Inspection and Test. Likewise, following the periodic inspection and testing of an existing installation, a Periodic Inspection Report, together with a schedule of inspections and a schedule of test results, is required to be given to the person ordering the inspection.

Sometimes the person ordering the work is not the user. It is necessary for the user (e.g. employer or householder) to have a copy of the inspection and test documentation. It is recommended that those providing documentation to the person ordering the work recommend that the forms be passed to the user including any purchaser of a domestic property.

1.3.2 Landlords and tenants

A landlord is required to provide a tenant with an electrical installation in good condition and repair. The landlord should maintain the installation in a condition suitable for the use intended, and ensure that repairs are undertaken by a competent person. A tenant has a duty to ensure that those parts of the installation that are his or her responsibility are maintained in a safe condition, and to ensure that repairs are carried out only by a competent person.

1.4 Alterations and additions

721 Every alteration or addition to an existing installation must comply with the Regulations and must not impair the safety of the existing installation. 130-07-01 721-01-01 721-01-02

743 When inspecting and testing an alteration or addition to an electrical installation, the existing installation must be inspected and tested so far as is necessary to ensure the safety of the alteration or addition,

including for example:

protective conductor continuity

earth fault loop impedance.

Whilst there is no obligation to inspect and test any part of the existing installation that does not affect and is not affected by the alteration or addition, observed departures are required to be noted in the comments box of Electrical Installation Certificates (single signature or multiple signature) and Minor Works Certificates. 743-01-02

1.5 Record keeping

Records of all checks, inspections and tests, including test results, should be kept throughout the working life of an electrical installation. This will enable deterioration to be identified. They can also be used as a management tool to ensure that maintenance checks are being carried out and to assess their effectiveness.

Section 2 — Initial verification

2.1 Purpose of initial verification

71 Initial verification, in the context of Regulation 711-01-01, is intended 711-01-01
741 to confirm that the installation complies with the designer's intentions 713-01-01
742 and has been constructed, inspected and tested in accordance with BS 7671.

This Section makes recommendations for the initial inspection and testing of electrical installations.

711 Chapter 71 of BS 7671 states the requirements for 'INITIAL VERIFICATION'. As far as reasonably practicable, an inspection shall be 712-01-02
712
713 carried out to verify:

(i) all fixed equipment and material is of the correct type and complies with applicable British Standards or acceptable equivalents

(ii) all parts of the fixed installation are correctly selected and erected

(iii) no part of the fixed installation is visibly damaged or otherwise defective.

Inspections 712

Inspection is an important element of inspection and testing, and is described in Section 2.6 of this Guidance Note.

Tests 713

The tests are described in Section 2.7 and 3.10 of this Guidance Note.

Results 741-01-01

The results of inspection and tests are to be recorded as appropriate. The Memorandum of Guidance on the Electricity at Work Regulations (EAW) recommends records of all maintenance including test results be kept throughout the life of an installation - see guidance on EAW Regulation 4(2). This can enable the condition of equipment and the effectiveness of maintenance to be monitored.

Relevant Criteria 713-01-01

The relevant criteria are, for the most part, the requirements of the Regulations for the particular inspection or test. The criteria are given in Sections 2 and 3 of this Guidance Note.

There will be some instances where the designer has specified requirements which are particular to the installation concerned. For example, the intended impedances may be different from those in BS 7671. In this case, the inspector should either ask for the design criteria or forward the test results to the designer for verification with the intended design. In the absence of such data the inspector should apply the requirements set out in BS 7671.

Verification

The responsibility for comparing inspection and test results with relevant criteria, as required by Regulation 713-01-01, lies with the party responsible for inspecting and testing the installation. This party, which may be the person carrying out the inspection and testing, should sign the inspection and testing box of the Electrical Installation Certificate or the declaration box of the Minor Electrical Installation Works Certificate. If the person carrying out the inspection and testing has also been responsible for the design and construction of the installation, he (or she) must also sign the design and construction boxes of the Electrical Installation Certificate, or make use of the single signature Electrical Installation Certificate.

2.2 Certificates

App 6 Appendix 6 of BS 7671 allows the use of three forms for the initial certification of a new installation or for an alteration or addition to an existing installation as follows:

multiple signature Electrical Installation Certificate

single signature Electrical Installation Certificate

Minor Electrical Installation Works Certificate.

Examples of typical forms are given in Section 5.

Multiple signature Electrical Installation Certificate

The multiple signature certificate allows different persons to sign for design, construction, inspection and testing, and allows two signatories for design where there is mutual responsibility. Where designers are responsible for identifiably separate parts of an installation, separate forms would be appropriate.

Single signature Electrical Installation Certificate

Where design, construction, inspection and testing are the responsibility of one person a certificate with a single signature may replace the multiple signature form.

(See the 'short form' of Section 5).

Minor Electrical Installation Works Certificate

This certificate is to be used only for minor works that do not include the provision of a new circuit, such as an additional socket-outlet or lighting point to an existing circuit.

2.3 Required information

311 BS 7671 requires that the following information shall be made available 711-01-02
312 to the person or persons carrying out the inspection and testing:

313 **413** *Assessment of general characteristics*

514 (i) the maximum demand, expressed in amperes per phase (after 311-01-01
711 diversity is taken into account)

(ii) the number and type of live conductors of the source of energy and of the circuits used in the installation 312-02-01

(iii) the type of earthing arrangement used by the installation and 312-03-01
any facilities provided by the supplier for the user. 313-01-01

(iv) the nominal voltage(s)

(v) the nature of the load current and supply frequency

(vi) the prospective fault current at the origin of the installation

(vii) the earth fault loop impedance (Ze) of that part of the system external to the installation

(viii) the suitability for the requirements of the installation, including the maximum demand

(ix) the type and rating of the overcurrent protective device acting at the origin of the installation.

Note:
These characteristics should also be available for safety services such as UPS and generators.

Diagrams, charts or tables

The information below regarding the basis of the design must be made available. This is also a non-specific requirement of the Health and Safety at Work etc Act: 514-09-01

(x) the type and composition of circuits, including points of utilisation, number and size of conductors and type of cable. This should include the Installation Method shown in Appendix 4 (paragraph 8) of BS 7671

(xi) the method used for compliance with the requirements for 413-01-01
protection against indirect contact and, where appropriate, the 413-02-04
conditions required for automatic disconnection

131 (xii) the information necessary for the identification of each device performing the functions of protection, isolation and switching, and its location

(xiii) any circuit or equipment vulnerable to a particular test.

2.4 Scope

It is essential that the inspector knows the extent of the installation to be inspected and any criteria regarding the limit of the inspection. This should be recorded on the Certificate.

2.5 Frequency of subsequent inspections

133
514
732

The proposed interval between periodic inspections is an element of the design, selection and erection of the installation. This interval is required to be noted on the Electrical Installation Certificate and on a notice to be fixed in a prominent position at or near the origin of the installation. 514-12-01

In the light of the inspector's knowledge of the installation, its use and environment he is required to recommend future intervals. The information and tables in Section 3 of this Guidance Note have been prepared to provide guidance. 133-03-01 732-01-01

2.6 Initial inspection

2.6.1 General procedure

711

Inspection, and where appropriate testing, should be carried out and recorded on suitable schedules progressively throughout the different stages of erection and before the installation is certified and put into service.

A model Schedule of Inspections is shown in Section 5.

2.6.2 Comments on individual items to be inspected

712

The inspection should include at least the checking of those items listed in Section 712 of BS 7671.

712-01-03 (i)

a *Connection of conductors* 526

Every connection between conductors and equipment/other conductors should provide durable electrical continuity and adequate mechanical strength. Requirements for the enclosure of and accessibility of connections must be considered. 513-01-01

712-01-03 (ii)

b *Identification of non-flexible cables and conductors*

Table 51A of BS 7671 provides a schedule of Colour Identification of each core of non-flexible cables and bare conductors. 514

It should be checked that each core or bare conductor is identified as necessary. Busbar and pole colour should also comply with Table 51A. 514-06-03

Where it is desired to indicate phase rotation, or a different function for cables of the same colour, numbered or lettered sleeves are permitted.

712-01-03 (ii)

c *Identification of flexible cables and cords*

Table 51B of BS 7671 provides a schedule of colour identification of cores of flexible cables and cords. It should be checked that each core is identified as necessary. The single colour green or the single colour Table 51B 514-07-01 514-07-02 yellow must not be used. Only the colour combination green-and-yellow is permitted and is only to be used for protective conductors. Where it is desired to indicate phase rotation or a different function for cables of the same colour, numbered or lettered sleeves are permitted.

712-01-03 (iii) *d* *Routing of cables*

Cables should be routed as appropriate out of harm's way, and where necessary, additionally protected against mechanical damage. 522

712-01-03 (iv) *e* *Cable selection*

Where practicable, the cable size should be assessed against the protective arrangement based upon information provided by the installation designer (where available). 131-06 523 524 525

Reference should be made, as appropriate, to Appendix 4 of BS 7671.

712-01-03 (v) *f* *Verification of polarity — single-pole device in a TN or TT system* 131-13

It must be verified that single-pole devices for protection or switching are installed in phase conductors only. 530

712-01-03 (vi) *g* *Accessories and equipment* 553

Correct connection (suitability, polarity etc) must be checked.

712-01-03 (xiv)

Table 55A of BS 7671 is a schedule of types of plug and socket-outlet available, the ratings, and the associated British Standards.

Particular attention should be paid to the requirements for cable couplers. 553-02-01

Bayonet lampholders B15 and B22 should comply with BS EN 61184 and be of temperature rating T2. 553-03-03

712-01-03 (vii) *h* *Selection and erection to minimise the spread of fire* 527-02 527-03

Fire barriers, suitable seals and/or protection against thermal effects should be provided if necessary to meet the requirements of BS 7671.

BS 7671 requires that each sealing arrangement be inspected to verify that it conforms with the manufacturer's erection instructions. This may be impossible without dismantling the system and it is essential, therefore, that inspection should be carried out at the appropriate stage of the work, and that this is recorded at the time for incorporation in the inspection and test documents. 527-04

712-01-03 (viii)(a) *i* *Protection against both direct and indirect contact*

SELV is the common method of providing protection against both direct and indirect contact. The many requirements include: 411-02

(a) the nominal voltage must not exceed 50 V a.c. or 120 V d.c.

(b) an isolated source e.g. a safety isolating transformer to BS 3535

(c) electrical separation from higher voltage systems

(d) no connection with earth by the SELV circuits

(e) SELV exposed-conductive-parts shall have no connection with earth, exposed-conductive-parts or protective conductors of other systems.

712-01-03 (viii)(b) *j* *Protection against direct contact* 412

Insulation

Although protection by insulation is the usual method, there are other methods of protection against direct contact. 412-02 471-04

Barriers or enclosures

Where live parts are protected by barriers or enclosures, these should be checked for adequacy and security. 412 03 471-05

Obstacles

Protection by obstacles provides protection only against unintentional contact. If this method is used the area shall be accessible only to skilled persons or to instructed persons under supervision. This method of protection is not to be used in some installations and locations of increased shock risk. See Part 6 of BS 7671. 412-04 471-06

Out of reach 412-05 471-07

Placing out of reach protects also against direct contact. Increased distances are necessary where long or bulky conducting objects are likely to be handled in the vicinity.

Bare live parts are permitted in an area accessible only to skilled persons, and the dimensions of passageways should be checked against the guidance in Appendix 3 of the Memorandum of Guidance on the Electricity at Work Regulations issued by the Health and Safety Executive.

PELV

The requirements for PELV are as for SELV except that the secondary circuits are earthed and exposed-conductive-parts may have connections with earth, exposed-conductive-parts or protective conductors of other systems. 471-14-01 471-14-02

712-01-03 (viii)(c) *k* *Protection against indirect contact* 413

The 'Methods of Protection against Indirect Contact' are classified in a number of sub-sections in BS 7671, and are:

(i) earthed equipotential bonding and automatic disconnection of supply 413-01-01

(ii) use of Class II equipment or equivalent insulation. Where equivalent insulation is used, the designer should be consulted for guidance.

(iii) non-conducting location

(iv) earth-free local equipotential bonding

(v) electrical separation.

Methods (ii) (iii) and (iv) are specialised protection systems. It is essential that inspection of such systems be carried out by persons competent in the discipline and having adequate information on the design of the system. For these specialised systems, the designer and client will advise of and agree, the necessary effective and continuing supervision.

(i) earthed equipotential bonding and automatic disconnection of supply (EEBADS) 413-02 471-08

The presence, correct sizing, labelling and connection of appropriate protective conductors must be confirmed as follows: 542

— earthing conductor 542-03

— main equipotential bonding conductors 413-02-02 547-02

— circuit protective conductors 543-03

— supplementary equipotential bonding conductors. 413-02-04 547-03

The earthing system must be determined, e.g. 312-03

— PME earth (TN-C-S system)

— TN-S earth

— earth electrode (TT system).

The earth impedance must be appropriate for the protective device i.e. RCD or overcurrent device. 413-02-04 413-02-08

Specialised systems

(ii) use of Class II equipment

This method of protection has very specialised requirements including continuing supervision to ensure the requirements continue to be met throughout the life of the installation.

(iii) non-conducting location

Protection by this method (e.g. television test area) should be applied only in a special situation under supervision of a suitably qualified person. 413-04 471-10

(iv) earth-free local equipotential bonding

This method is applied in a special situation which is earth-free, under effective supervision, and where specified by a suitably qualified electrical engineer. A warning notice complying with Section 514 must be fixed in a prominent position adjacent to every point of access to the location concerned. This method is sometimes combined with 'Electrical Separation'. 413-05 471-11 514-13-02

Inspection should verify that no item is earthed within the area and that no earthed services or conductors enter or traverse the area, including the floor and ceiling. Inspection should confirm the designer's intentions and whether or not they have been achieved.

(v) electrical separation

This method may be applied to the supply of an individual item of 413-06
equipment through an isolating transformer to BS 3535. In special 471-12
situations, under effective supervision, when specified by a suitably 514-13-02
qualified engineer, it may be used to supply several items of equipment from the same source.

712-01-03 (ix) *l* *Prevention of mutual detrimental influence*

The requirements are stated in Regulation 131-11-01, Chapter 33: 131-11
Compatibility, and in Section 515: Mutual Detrimental Influence. 33
Another aspect which should be taken into consideration during 515
inspection is Section 528: Proximity to other Services. 528

712-01-03 (x) *m* *Isolating and switching devices*

BS EN 60947-1 (Specification for low voltage switchgear and controlgear - General rules) defines standard utilization categories which allow for conditions of service use and the switching duty to be expected.

All switch utilization categories must be appropriate for the nature of the load — see Table 2.1. Checking of utilization category may need to be carried out during construction, if the label is obscured during erection. Guidance Note 2 (Isolation & Switching) provides more comprehensive guidance on this subject and should be consulted and its contents taken into account.

Switchgear to BS EN 60947-1 if suitable for isolation will be marked with the symbol:

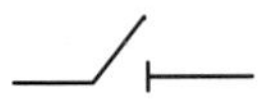

This may be endorsed with a symbol advising of function, e.g. for a switch disconnector:

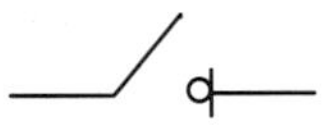

TABLE 2.1
Examples of utilization categories for alternating current installations

Utilization category		Typical applications
Frequent operation	Infrequent operation	
AC-20A	AC-20B	- Connecting and disconnecting under no-load conditions
AC-21A	AC-21B	- Switching of resistive loads including moderate overloads
AC-22A	AC-22B	- Switching of mixed resistive and inductive loads, including moderate overloads
AC-23A	AC-23B	- Switching of motor loads or other highly inductive loads

An isolation exercise should be carried out to check that effective isolation can be achieved. This should include, where appropriate, locking-off and inspection or testing to verify that the circuit is dead and no other source of supply is present.

712-01-03 (xi) *n* *Presence of undervoltage protective devices*

131-08-01
451-01-01

Suitable precautions should be in place where a reduction in voltage, or loss and subsequent restoration of voltage, could cause danger. Normally such a requirement concerns only motor circuits; if it is required it will have been specified by the designer.

712-01-03 (xii) *o* *Protective devices* 413
43

The choice and setting of each protective device including monitoring devices should be compared with the design.

712-01-03 (xiii) *p* *Labelling of protective devices, switches and terminals* 514-08

A protective device must be arranged and identified so that the circuit protected may be easily recognised.

712-01-03 (xiv) *q* *Selection of equipment and protective measures appropriate to external influences* 131-05
512-06
522

Equipment must be selected with regard to its suitability for the environment — ambient temperature, heat, water, foreign bodies, corrosion, impact, vibration, flora, fauna, radiation, building use and structure.

712-01-03 (xv) *r* *Adequacy of access to switchgear and equipment* 131-12
513

Every piece of equipment which requires operation or attention by a person must be so installed that adequate and safe means of access and working space are afforded.

712-01-03 (xvi) s *Presence of danger notices and other warning notices* 514

Suitable warning notices, suitably located, are required to be installed to give warning of:

Voltage 514-10

- where a nominal voltage exceeding 230 volts exists within an item of equipment or enclosure and where the presence of such a voltage would not normally be expected
- where a nominal voltage exceeding 230 volts exists between simultaneously accessible terminals or other fixed live parts
- where different nominal voltages exist.

Isolation 514-11

where live parts are not capable of being isolated by a single device.

Periodic Inspection and Testing 514-12

the wording of the notice is given in Regulation 514-12-01.

RCDs 514-12

the wording of the notice is given in Regulation 514-12-02.

Earthing and bonding conductors 514-13

- the requirements for the label and its wording are given in Regulation 514-13-01
- the wording of the notice is given in Regulation 514-13-02.

712-01-03 (xvii) *t* *Presence of diagrams, instructions and similar information* 514-09

A schedule within or adjacent to the distribution board is sufficient for simple installations.

712-01-03 (xviii) *u* *Erection methods* Part 5

Chapter 52 contains detailed requirements on selection and erection. Fixings of switchgear, cables, conduit, fittings, etc must be adequate for the environment.

2.6.3 Inspection checklist

(This checklist may also be used when carrying out periodic inspections.)

Listed below are requirements to be checked when carrying out an installation inspection. The list is not exhaustive.

General

1. Complies with requirements (i) - (iii) in Section 2.1 (132-01-01, 133-01-01)
2. Accessible for operation, inspection and maintenance (513-01-01)
3. Suitable for local atmosphere and ambient temperature (Chap 52) (Installations in potentially explosive atmospheres are outside the scope of BS 7671, see BS EN 60079 and BS EN 50014)
4. Circuits to be separate (no borrowed neutrals) (314-01-04)
5. Circuits to be identified (neutral and protective conductors in same sequence as phase conductors) (514-01-02, 514-08-01)
6. Protective devices adequate for intended purpose (Chap 53)
7. Disconnection times likely to be met by installed protective devices (Chap 41)
8. More than one socket provided for convenience (553-01-07)
9. All circuits suitably identified (514-09)
10. Suitable main switch provided (Chap 46)
11. Supplies to any safety services suitably installed e.g. Fire Alarms BS 5839
12. Environmental IP requirements accounted for (BS EN 60529)
13. Means of isolation suitably labelled (514-01-01, 537-02-09)
14. Provision for disconnecting the neutral (460-01-06)
15. Main switches to single-phase installations, intended for use by an ordinary person, e.g. domestic, shop, office premises, to be double-pole (476-01-03)
16. RCDs provided where required (471-08, 471-16)
17. Discrimination between RCDs considered (314-01-02, 531-02-09)
18. Main earthing terminal provided (542-04-01) readily accessible and identified (514-13-01)
19. Provision for disconnecting earthing conductor (542-04-02)
20. Correct cable glands and gland-plates used (BS 6121)
21. Cables used comply with British or Harmonized Standards (Appendix 4 of the Regulations, 521-01-01)
22. Conductors correctly identified (Sect 514)
23. Earth tail pots installed where required on mineral insulated cables (133-01-04)
24. Non-conductive finishes on enclosures removed to ensure good electrical connection and if necessary made good after connecting (526-01-01)
25. Adequately rated distribution boards (BS 5486, or BS EN 60439 may require derating)
26. Correct fuses or circuit-breakers installed (Sect 531 and Sect 533)
27. All connections secure (133-01-01)
28. Consideration paid to electromagnetic effects and electromechanical stresses (Chap 52)

29. Overcurrent protection provided where applicable (Sect 473 and Sect 533)
30. Segregation of circuits (Sect 515 and Sect 528)
31. Retest notice provided (514-12-01)
32. Sealing of the wiring system including fire barriers (527-02).

Switchgear

1. Suitable for the purpose intended (Chap 53)
2. Meets requirements of BS EN 61008, BS EN 61009, BS EN 60947-2, BS EN 60898 or BS EN 60439 where applicable, or equivalent standards (511)
3. Securely fixed (133-01-01) and suitably labelled (514-01)
4. Non-conductive finishes on switchgear removed at protective conductor connections and if necessary made good after connecting (526-01-01)
5. Suitable cable glands and gland plates used (526-01)
6. Correctly earthed (Chap 54)
7. Conditions likely to be encountered taken account of i.e. suitable for the foreseen environment (Sect 522)
8. Correct IP rating applied (BS EN 60529)
9. Suitable as means of isolation, where applicable (537-02)
10. Complies with the requirements for locations containing a bath or shower (Sect 601)
11. Need for isolation, mechanical maintenance, emergency and functional switching met (Sect 537)
12. Fireman's switch provided where required (476-03, 537-04-06)
13. Switchgear suitably coloured where necessary (537-04-04)
14. All connections secure (Sect 526)
15. Cables correctly terminated and identified (Sect 514 and Sect 526)
16. No sharp edges on cable entries, screw heads etc which could cause damage to cables (522-08)
17. All covers and equipment in place and secure (Sect 510)
18. Adequate access and working space (131-12-01 and Sect 513).

Installation check lists

Wiring accessories

General (applicable to each type of accessory)

1. Complies with BS 5733, BS 6220 or other appropriate standard (Sect 511)
2. Box or other enclosure securely fixed (133-01-01)
3. Metal box or other enclosure earthed (471-08-08 and Chap 54)
4. Edge of flush boxes not projecting above wall surface (526-03)
5. No sharp edges on cable entries, screw heads, etc which could cause damage to cables (522-08)
6. Non-sheathed cables, and cores of cable from which sheath has been removed, not exposed outside the enclosure (526-03)
7. Conductors correctly identified (514-06, 514-07)
8. Bare protective conductors sleeved green-and-yellow (514-03, 543-03-02)
9. Terminals tight and containing all strands of the conductors (Sect 526)
10. Cord grip correctly used or clips fitted to cables to prevent strain on the terminals (522-08-05)
11. Adequate current rating (132-01-04)
12. Suitable for the conditions likely to be encountered (Sect 522).

Lighting controls

1. Light switches comply with BS 3676 or BS 5518 (Sect 511)
2. Suitably located (Sect 512)
3. Single-pole switches connected in phase conductors only (131-13-01)
4. Correct colour coding or marking of conductors (514-06-01)
5. Earthing of exposed metalwork, e.g. metal switchplate (Chap 54)
6. Complies with the requirements for locations containing a bath or shower (Sect 601)
7. Adequate current rating (132-01-04)
8. Suitable for inductive circuits or de-rated where necessary (512-02)
9. Switch labelled to indicate purpose, where this is not obvious (514-01)
10. Track systems comply with BS EN 60570(521-06)
11. Appropriate controls suitable for the luminaires (553-04-01).

Lighting points

1. Correctly terminated in a suitable accessory or fitting (553-04-01)
2. Ceiling rose complies with BS 67 (553-04-01)
3. Not more than one flex unless designed for multiple pendants (553-04-03)
4. Flex support devices used (553-04-04)
5. Switch wires identified (514-06-01)
6. Holes in ceiling above rose made good to prevent spread of fire (527-02-01)
7. Not connected to a supply exceeding 250 V (553-04-02)

8. Suitable for the mass suspended (554-01-01)
9. Lampholders to BS 7895, BS EN 60238 or BS EN 61184
10. Luminaire couplers comply with BS 6972 or BS 7001 (553-04-01).

Socket-outlets

1. Complies with BS 196, BS 546, BS 1363, BS 4343, BS EN 60309-2 (553-01) and shuttered for household and similar installations (553-01-04)
2. Mounting height above the floor or working surface suitable (553-01-06)
3. Correct polarity (131-13-01 and 713-09)
4. Not installed in a bathroom or shower room unless shaver or SELV (601-08-01)
5. Outside the zones in a room other than a bathroom or shower room and RCD protected (601-08-02)
6. Controlled by a switch, where the supply is direct current (537-05-05)
7. Protected where mounted in a floor (Sect 522)
8. Not used to supply a water heater having uninsulated elements (554-05-03)
9. Circuit protective conductor connected directly to the earthing terminal of the socket-outlet, on a sheathed wiring installation (543-02-07)
10. Earthing tail from the earthed metal box, on a conduit installation to the earthing terminal of the socket-outlet (543-02-07).

Joint box

1. Joints accessible for inspection (526-04)
2. Joints protected against mechanical damage (526-03-01)
3. All conductors correctly connected (526-01)
4. Complies with BS 5733 or BS 1363-4 (553-04-01vi).

Fused connection unit

1. Complies with the requirements for locations containing a bath or shower (Sect 601)
2. Correct rating and fuse (533-01).

Cooker control unit

1. Sited to one side and low enough for accessibility and to prevent flexes trailing across radiant plates (131-12-01, 476-03-04)
2. Cable to cooker fixed to prevent strain on connections (526-01)

Conduits

General

1. Securely fixed, covers in place and adequately protected against mechanical damage (522-08)
2. Inspection fittings accessible (522-08-02)
3. Number of cables for easy draw not exceeded (522-08-01)
4. Solid elbows and tees used only as permitted (522-08-01 and 522-08-03)
5. Ends of conduit reamed and bushed (522-08)
6. Adequate boxes (522-08)
7. Unused entries blanked off where necessary (412-03)
8. Not containing unsuitable non-electrical pipes or tubes (Sect 511)
9. Provided with drainage holes and gaskets as necessary (522-03)
10. Radius of bends such that cables are not damaged (522-08-03)
11. Joints, scratches etc in metal conduit protected by painting (133-01-01, 522-05).

Rigid metal conduit

1. Complies with BS 31, or BS 4568 Parts 1 and 2 (Sect 511)
2. Connected to the main earth terminal (413-02-06)
3. Phase and neutral cables contained in the same conduit (521-02)
4. Conduit suitable for damp and corrosive situations (522-03 and 522-05)
5. Maximum span between buildings without intermediate support (522-08 and see Guidance Note 1 and On-Site Guide)

Rigid non-metallic conduit

1. Complies with BS 4607, BS EN 60423 and BS EN 50086-2-1 (521-04)
2. Ambient and working temperatures within permitted limits (522-01 and 522-02)
3. Provision for expansion and contraction (522-08)
4. Boxes and fixings suitable for mass of luminaire suspended at expected temperature (522-08, 554-01-01).

Flexible metal conduit

1. Complies with BS 731-1, BS EN 60423 and BS EN 50086-1 (521-04-01)
2. Separate protective conductor provided (543-02-01)
3. Adequately supported and terminated (522-08).

Trunking

General

1. Complies with BS 4678 or BS EN 50085-1 (521-05-01)
2. Securely fixed and adequately protected against mechanical damage (522-08)
3. Selected, erected and routed so that no damage is caused by ingress of water (522-03)
4. Proximity to non-electrical services (528-02)

5. Internal sealing provided where necessary (527-02)
6. Holes surrounding trunking made good (527-02)
7. Band I circuits partitioned from Band II circuits or insulated for the highest voltage present (528-01)
8. Circuits partitioned from Band I circuits or wired in mineral-insulated metal-sheathed cables (528-01)
9. Common outlets for Band I and Band II provided with screens, barriers or partitions (528-01)
10. Cables supported for vertical runs (522-08).

Metal trunking

1. Phase and neutral cables contained in the same metal trunking (521-02)
2. Protected against damp or corrosion (522-03 and 522-05)
3. Earthed (413-02)
4. Joints mechanically sound, and of adequate continuity with links fitted (543-02).

Insulated Cables

Non-flexible cables

1. Correct type (521-01)
2. Correct current rating (523-01)
3. Protected against mechanical damage and abrasion (522-08)
4. Cables suitable for high or low ambient temperature as necessary (522-01)
5. Non-sheathed cables protected by enclosure in conduit, duct or trunking (521-07)
6. Sheathed cables routed in allowed zones or mechanical protection provided (522-06).
7. Where exposed to direct sunlight, of a suitable type (522-11)
8. Not run in lift shaft unless part of the lift installation and of the permitted type (528-02-06 and BS 5655)
9. Correctly selected and installed for use e.g. buried (522-06-03)
10. Correctly selected and installed for use on exterior walls etc (Sect 522)
11. Correctly selected and installed for use overhead (521-01-03 and 522-08)
12. Internal radii of bends in accordance with (relevant BS and 522-08)
13. Correctly supported (522-08-04)
14. Not exposed to water etc unless suitable for such exposure (522-03)
15. Metal sheaths and armour earthed (543-02)
16. Identified at terminations (514-06)
17. Joints and connections electrically and mechanically sound and adequately insulated (526-01 and 526-02)
18. All wires securely contained in terminals etc without strain (Sect 526)
19. Enclosure of terminals (Sect 526)
20. Glands correctly selected and fitted with shrouds and supplementary earth tags as necessary (526-01)

21. Joints and connections mechanically sound and accessible for inspection, except as permitted otherwise (526-04)
22. Earthed concentric wiring including cne cables to be used only as permitted (546-02 and Electricity Supply Regulations 1988 (as amended)).

Flexible cables and cords

1. Correct type (521-01-01)
2. Correct current rating (Sect 523)
3. Protected where exposed to mechanical damage (522-06 and 522-08)
4. Suitably sheathed where exposed to contact with water (522-03) and corrosive substances (522-05)
5. Protected where used for final connections to fixed apparatus etc (521-07 and 526-03-03)
6. Selected for resistance to damage by heat (522-02)
7. Segregation of Band I and Band II circuits (BS 6701 and Sect 528)
8. Fire alarm and emergency lighting circuits segregated (BS 5839, BS 5266 and Section 528)
9. Cores correctly identified throughout (514-07)
10. Prohibited core colours not used (514-07-02)
11. Joints to be made using cable couplers (526-02-01)
12. Where used as fixed wiring relevant requirements met (521-01-04)
13. Final connections to portable equipment a convenient length and connected as stated (553-01-07)
14. Final connections to other current-using equipment properly secured or arranged to prevent strain on connections (Sect 554)
15. Mass supported by pendants not exceeding values stated (554-01).

Protective conductors

1. Cables incorporating protective conductors comply with the relevant BS (Sect 511)
2. Joints in metal conduit, duct or trunking comply with Regulations (543-03)
3. Flexible conduit to be supplemented by a protective conductor (543-02-01)
4. Minimum cross-sectional area of copper conductors (543-01)
5. Copper conductors, other than strip, of 6 mm^2 or less protected by insulation (543-03)
6. Circuit protective conductor at termination of sheathed cables insulated with sleeving (543-03-02)
7. Bare circuit protective conductor protected against mechanical damage and corrosion (542-03 and 543-03-01)
8. Insulation, sleeving and terminations identified by colour combination green-and-yellow (514-03-01)
9. Joints sound (526-01)
10. Main and supplementary bonding conductors of correct size (Sect 547)

11. Separate circuit protective conductors not less than 4 mm^2 if not protected against mechanical damage (543-01-01).

Enclosures

General

1. Suitable degree of protection (IP Code in BS EN 60529) appropriate to external influences (412-03, Sect 522 and Part 6).

2.7 Initial testing

Introduction to test methods

The test methods described in this section are the preferred test methods to be used; other appropriate test methods are not precluded.

2.7.1 Initial testing

713 The test results must be recorded on the Schedules of Test Results and compared with relevant criteria. 713-01-01

For example, relevant criteria for earth fault loop impedance may be provided by the designer, obtained as described in Section 2.7.14 or, where appropriate, obtained from Appendix 2 of this Guidance Note.

A model Schedule of Test Result is shown in Section 5.

2.7.2 Electrical Installation Certificate

741
742 Section 741 of BS 7671 requires that, upon completion of the verification of a new installation, or changes to an existing installation, an Electrical Installation Certificate based on the model given in Appendix 6 of BS 7671 shall be provided. Section 742 requires that:

(i) the Electrical Installation Certificate be accompanied by a Schedule of Inspections and a Schedule of Test Results. These schedules shall be based upon the models given in Appendix 6 of BS 7671 742-01-01

(ii) the Schedule of Test Results shall identify every circuit, including its related protective device(s), and shall record the results of the appropriate tests and measurements detailed in Chapter 71 742-01-02

(iii) the Electrical Installation Certificate shall be signed by a competent person or persons stating that to the best of their knowledge and belief the installation has been designed, constructed, inspected and tested in accordance with BS 7671, any permissible deviations being listed 741-01-04 742-01-03

(iv) any defects or omissions revealed by the Inspector shall be made good and inspected and tested again before the Electrical Installation Certificate is issued. 742-01-04

2.7.3 Model forms

Typical forms for use when carrying out inspection and testing are included in Section 5 of this Guidance Note.

2.7.4 The sequence of tests

713 Initial tests should be carried out in the following sequence: 713-01-01

(a) continuity of protective conductors, including main and supplementary bonding (2.7.5);

(b) continuity of ring final circuit conductors (2.7.6);

(c) insulation resistance (2.7.7);

(d) site applied insulation (2.7.8);

(e) protection by separation of circuits (2.7.9);

(f) protection by barriers or enclosures provided during erection (2.7.10);

(g) insulation of non-conducting floors and walls (2.7.11);

(h) polarity (2.7.12);

(i) earth electrode resistance (2.7.13);

(j) earth fault loop impedance (2.7.14);

(k) prospective fault current (2.7.15);

(l) functional testing (2.7.16).

2.7.5 Continuity of protective conductors including main and supplementary bonding

Regulation 471-08-08 requires that installations which provide protection against indirect contact using EEBADS must have a circuit protective conductor run to and terminated at each point in the wiring and at each accessory. An exception is made for a lampholder having no exposed-conductive-parts and suspended from such a point. 471-08-08

Test methods 1 and 2 are alternative ways of testing the continuity of protective conductors.

Every protective conductor, including circuit protective conductors, the earthing conductor and main and supplementary equipotential bonding conductors, should be tested to verify that the conductors are electrically sound and correctly connected. 713-02-01

Test method 1 detailed below, as well as checking the continuity of the protective conductor, also measures (R_1+R_2) which, when added to the external impedance (Z_e), enables the earth-fault loop impedance (Z_S) to be checked against the design, see Section 2.7.14.

Note 1
(R_1 + R_2) is the sum of the resistance of the phase conductor R_1 and the circuit protective conductor R_2.

Note 2
The reading may be affected by parallel paths through exposed-conductive-parts and/or extraneous-conductive-parts.

It should also be recognised that test methods 1 and 2 can only be applied simply to an 'all insulated' installation. Installations incorporating steel conduit, steel trunking, micc and thermoplastic/swa cables will introduce parallel paths to protective conductors. Similarly, luminaires fitted in grid ceilings and suspended from steel structures in buildings will create parallel paths.

In such situations, unless a plug and socket arrangement has been incorporated in the lighting system by the designer, the (R_1 + R_2) test will need to be carried out prior to fixing accessories and bonding straps to the metal enclosures and finally connecting protective conductors to luminaires. Under these circumstances some of the requirements may have to be visually inspected after the test has been completed. This consideration requires tests to be performed during the erection of an installation, in addition to tests at the completion stage.

Instrument — Use a low-resistance ohmmeter for these tests. Refer to Section 4.3

The resistance readings obtained include the resistance of the test leads. The resistance of the test leads should be measured and deducted from all resistance readings obtained unless the instrument can auto-null.

Test method 1

Connect the phase conductor to the protective conductor at the distribution board or consumer unit so as to include all the circuit. Then test between phase and earth terminals at each outlet in the circuit. The measurement at the circuit's extremity should be recorded on the Schedule of Test Results as the value of (R_1 + R_2) for the circuit under test.

See Fig 1a for Test method connections.

Fig 1a: Connections for testing continuity of protective conductors Method 1

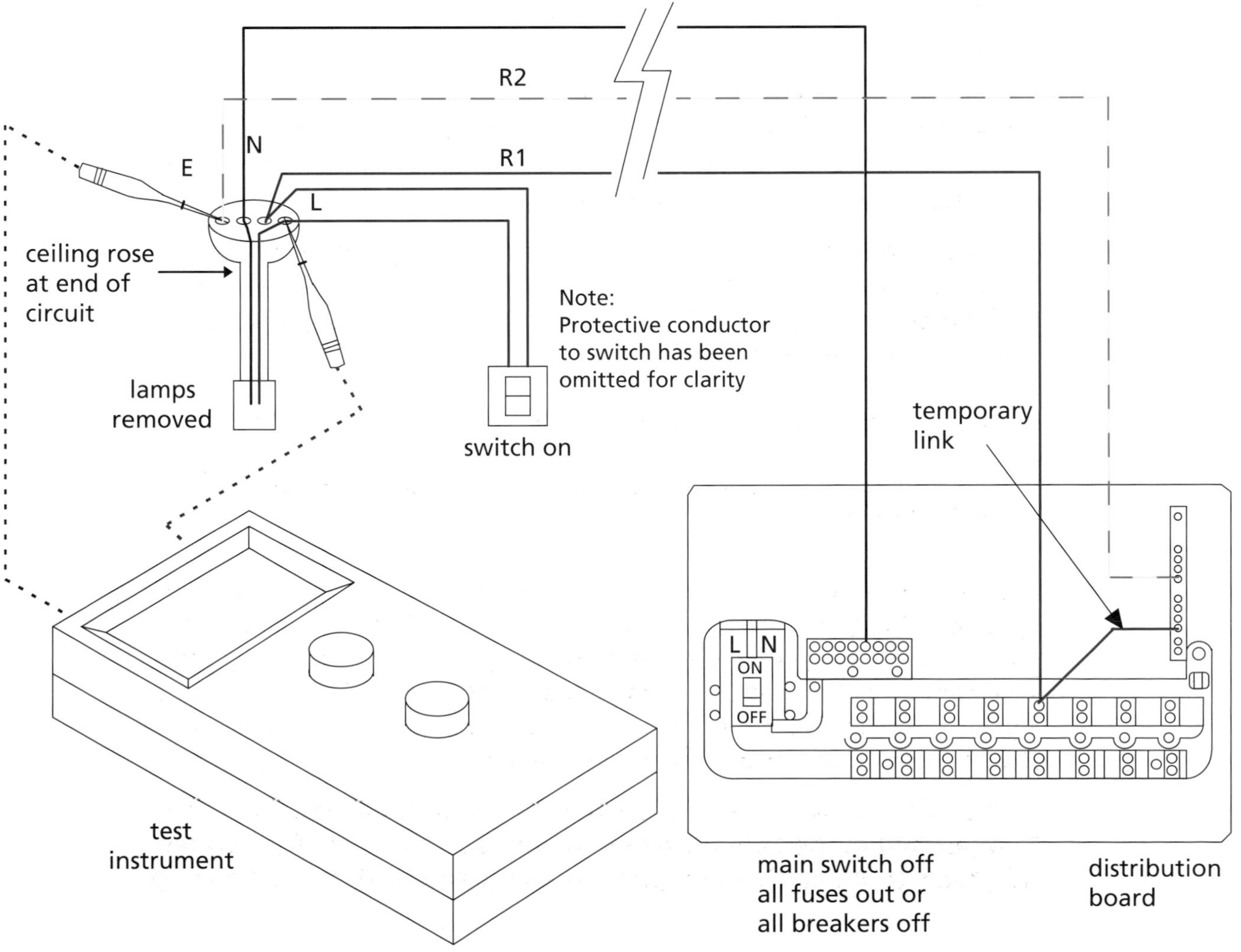

Test method 2

Connect one terminal of the continuity tester to the installation main earthing terminal, and with a test lead from the other terminal make contact with the protective conductors at various points on the circuit, e.g. luminaires, switches, spur outlets etc.

Bonding conductor continuity can be checked using this test method.

One end of the bonding conductor and any intermediate connections with services may need to be disconnected to avoid parallel paths.

Where indirect shock protection is provided by limiting the resistance of the circuit protective conductor to that given in Table 41C of BS 7671 (i.e. the 'alternative method'), it will be necessary to make a separate measurement of the resistance of the circuit protective conductor (R_2) using Test method 2. 413-02 12

This resistance R_2 is required to be recorded on the Schedule of Test Results where the 'alternative method' is used.

Fig 1b: Connections for testing continuity of protective conductors Method 2

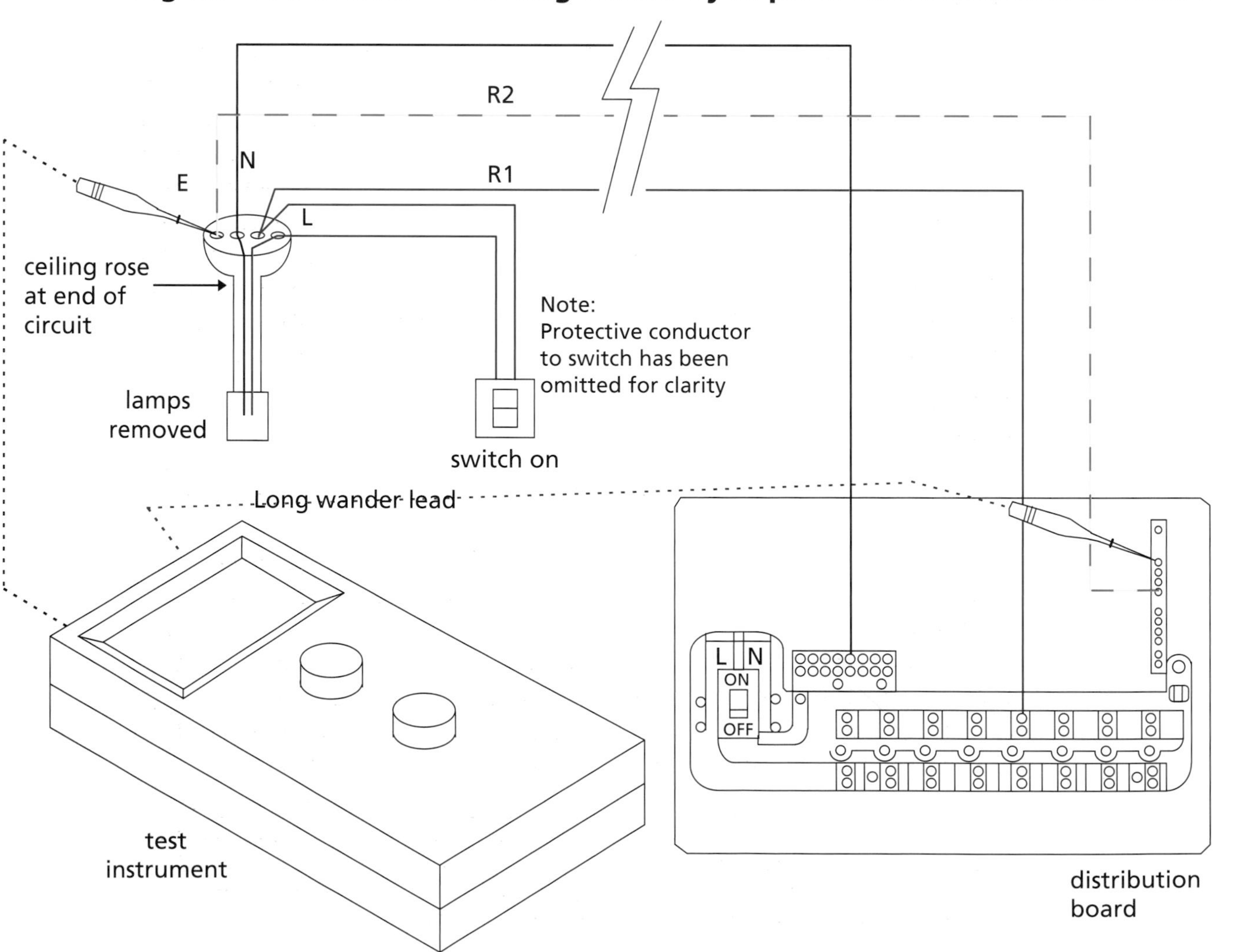

Where metallic enclosures have been used as the protective conductors, e.g. conduit, trunking, steel-wire armouring etc the following procedure should be followed:

(i) inspect the enclosure along its length for soundness of construction

(ii) perform the standard ohmmeter test using the appropriate test method described above.

Instrument: Use a low-resistance ohmmeter for this test.

2.7.6 Continuity of ring final circuit conductors

713 A three-step test is required to verify the continuity of the phase, neutral and protective conductors and correct wiring of every ring final circuit. The test results show if the ring has been inter-connected to create an apparently continuous ring circuit which is in fact broken. 713-03-01

Instrument: Use a low-resistance ohmmeter for these tests. Refer to Section 4.3

Step 1:

The phase, neutral and protective conductors are identified and the end-to-end resistance of each is measured separately (see Fig 2a). These resistances are r_1, r_n and r_2 respectively. A finite reading confirms that there is no open circuit on the ring conductors under test. The resistance values obtained should be the same (within 0.05 ohm) if the conductors are the same size. If the protective conductor has a reduced csa the resistance r_2 of the protective conductor loop will be proportionally higher than that of the phase or neutral loop e.g. 1.67 times for 2.5/1.5 mm^2 cable. If these relationships are not achieved then either the conductors are incorrectly identified or there is something wrong at one or more of the accessories.

Step 2:

The phase and neutral conductors are then connected together so that the outgoing phase conductor is connected to the returning neutral conductor and vice-versa (see Fig 2b). The resistance between phase and neutral conductors is measured at each socket-outlet. The readings at each of the sockets wired into the ring will be substantially the same and the value will be approximately one quarter of the resistance of the phase plus the neutral loop resistances, i.e. $(r_1 + r_n)/4$. Any sockets wired as spurs will give a higher resistance value due to the resistance of the spur conductors.

Note: Where single-core cables are used, care should be taken to verify that the phase and neutral conductors of opposite ends of the ring circuit are connected together. An error in this respect will be apparent from the readings taken at the socket-outlets, progressively increasing in value as readings are taken towards the midpoint of the ring, then decreasing again towards the other end of the ring.

Step 3:

The above step is then repeated but with the phase and cpc cross-connected (see Fig 2c). The resistance between phase and earth is measured at each socket-outlet. The readings obtained at each of the sockets wired into the ring will be substantially the same and the value will be approximately one quarter of the resistance of the phase plus cpc loop resistances, i.e. $(r_1 + r_2)/4$. As before, a higher resistance value will be recorded at any sockets wired as spurs. The highest value recorded represents the maximum $(R_1 + R_2)$ of the circuit and is recorded on the Schedule of Test Results. The value can be used to determine the earth loop impedance (Z_S) of the circuit to verify compliance with the loop impedance requirements of the Regulations (See Section 2.7.14).

This sequence of tests also verifies the polarity of each socket, except that if the testing has been carried out at the terminals on the reverse of the accessories a visual inspection is required to confirm correct polarity connections.

Fig 2: Connections for testing continuity of ring final circuit conductors

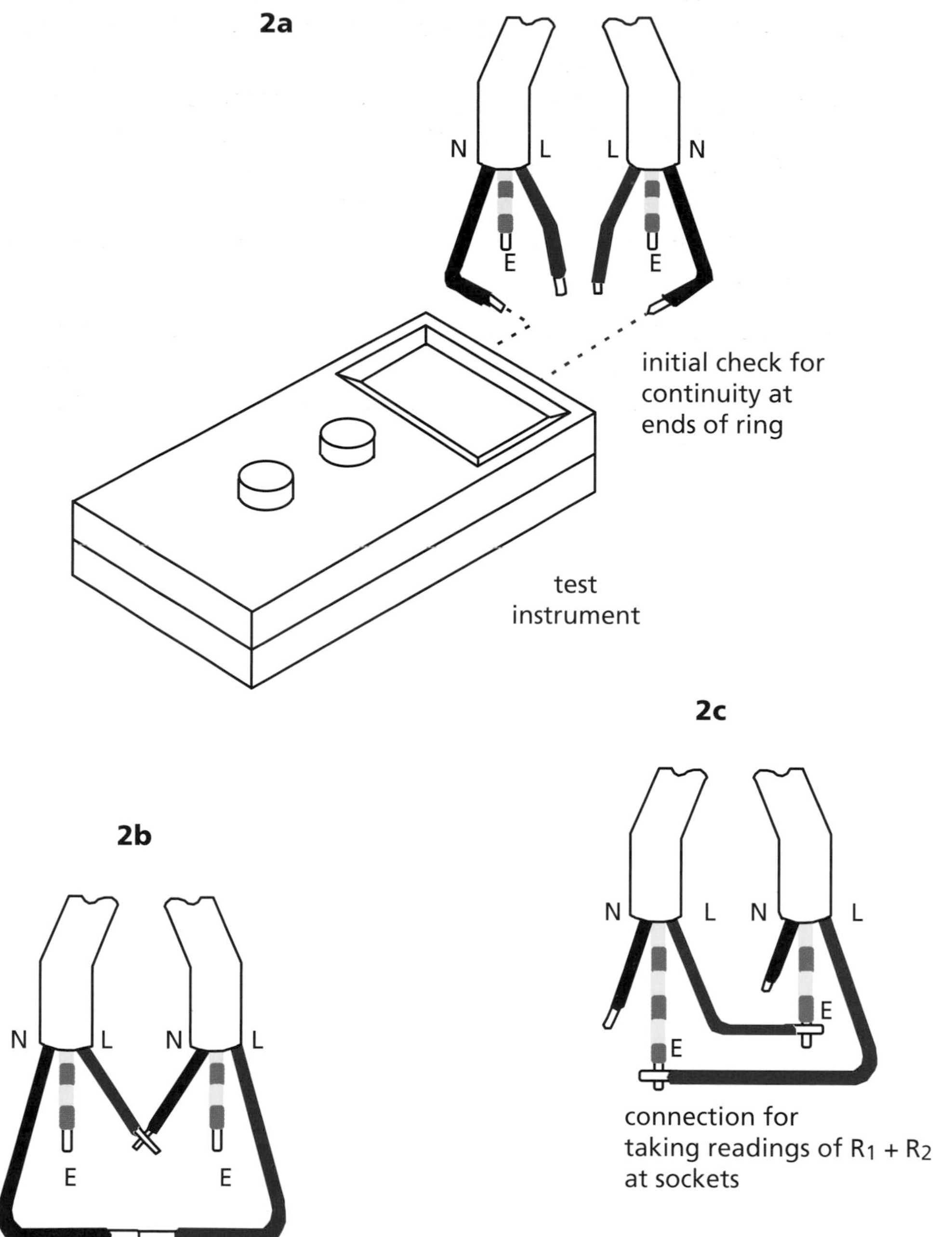

2.7.7 Insulation resistance

713 These tests are to verify that for compliance with BS 7671 the insulation of conductors and electrical accessories and equipment is satisfactory and that live conductors or protective conductors are not short-circuited, or show a low insulation resistance (which would indicate defective insulation). 713-04

Before testing check that:

(i) pilot or indicator lamps, and capacitors are disconnected from circuits to avoid an inaccurate test value being obtained

(ii) if Test 1 is to be used voltage-sensitive electronic equipment such as dimmer switches, touch switches, delay timers, power controllers, electronic starters for fluorescent lamps, emergency lighting, RCDs, etc are disconnected so that they are not subjected to the test voltage

(iii) there is no electrical connection between any phase or neutral conductor and earth.

If circuits contain voltage sensitive devices, Test 2, from protective earth to (phase and neutral) connected together, may be appropriate if vulnerable equipment is not to be disconnected. Further precautions may also be necessary to avoid damage to some electronic devices, and it may be necessary to consult the manufacturer of the equipment to identify necessary precautions.

Instrument: Use an insulation resistance tester for these tests. Refer to Section 4.4

Insulation resistance tests should be carried out using the appropriate d.c. test voltage specified in Table 71A of BS 7671. The installation will be deemed to conform with the Regulations if the main switchboard, and each distribution circuit tested separately with all its final circuits connected, but with current-using equipment disconnected, has an insulation resistance not less than that specified in Table 71A, which is reproduced here as Table 2.2.

Simple installations that contain no distribution circuits should be tested as a whole.

The tests should be carried out with the main switch off, all fuses in place, switches and circuit-breakers closed, lamps removed, and fluorescent and discharge luminaires and other equipment disconnected. Where the removal of lamps and/or the disconnection of current-using equipment is impracticable, the local switches controlling such lamps and/or equipment should be open.

TABLE 2.2
Minimum values of insulation resistance

Circuit nominal voltage (V)	Test Voltage d.c. (V)	Minimum insulation resistance (MΩ)
SELV and PELV	250	0.25
Up to and including 500 V with the exception of SELV and PELV but including FELV	500	0.5
Above 500 V	1000	1.0

To perform the test in a complex installation it may need to be sub-divided into its component parts.

Although an insulation resistance value of not less than 0.5 megohm complies with the Regulations, where an insulation resistance of less than 2 megohms is recorded, the possibility of a latent defect exists. In these circumstances, each circuit should be tested separately. This will help identify (i) whether one particular circuit in the installation has a lower insulation resistance value possibly indicating a latent defect that should be rectified or (ii) whether the low insulation resistance represents, for example, the summation of individual circuit insulation resistance and as such may not be a cause for concern.

Test 1, Insulation resistance between live conductors

Test between the live conductors at the appropriate distribution board.

Resistance readings obtained should be not less than the minimum values referred to in Table 2.2.

See Fig 3a for Test method connections.

Test 2, Insulation resistance to earth

Single-phase

Test between the phase and neutral conductors connected together and earth at the appropriate distribution board, or for circuits/equipment not vulnerable to insulation resistance testing, phase and neutral separately to earth.

For circuits containing two-way switching or two-way and intermediate switching the switches must be operated one at a time and the circuits subjected to additional insulation resistance tests.

Three-phase

Test to earth from all live conductors (including the neutral) connected together, or for circuits/equipment not vulnerable to insulation resistance tests, each live conductor separately to earth.

Where a low reading is obtained (less than 2 MΩ) it may be necessary to test each conductor separately to earth, after ensuring that all equipment is disconnected.

Resistance readings obtained should be not less than the minimum values referred to in Table 2.2.

See Fig 3b for Test method connections.

Fig 3a: Insulation resistance tests between live conductors of a circuit

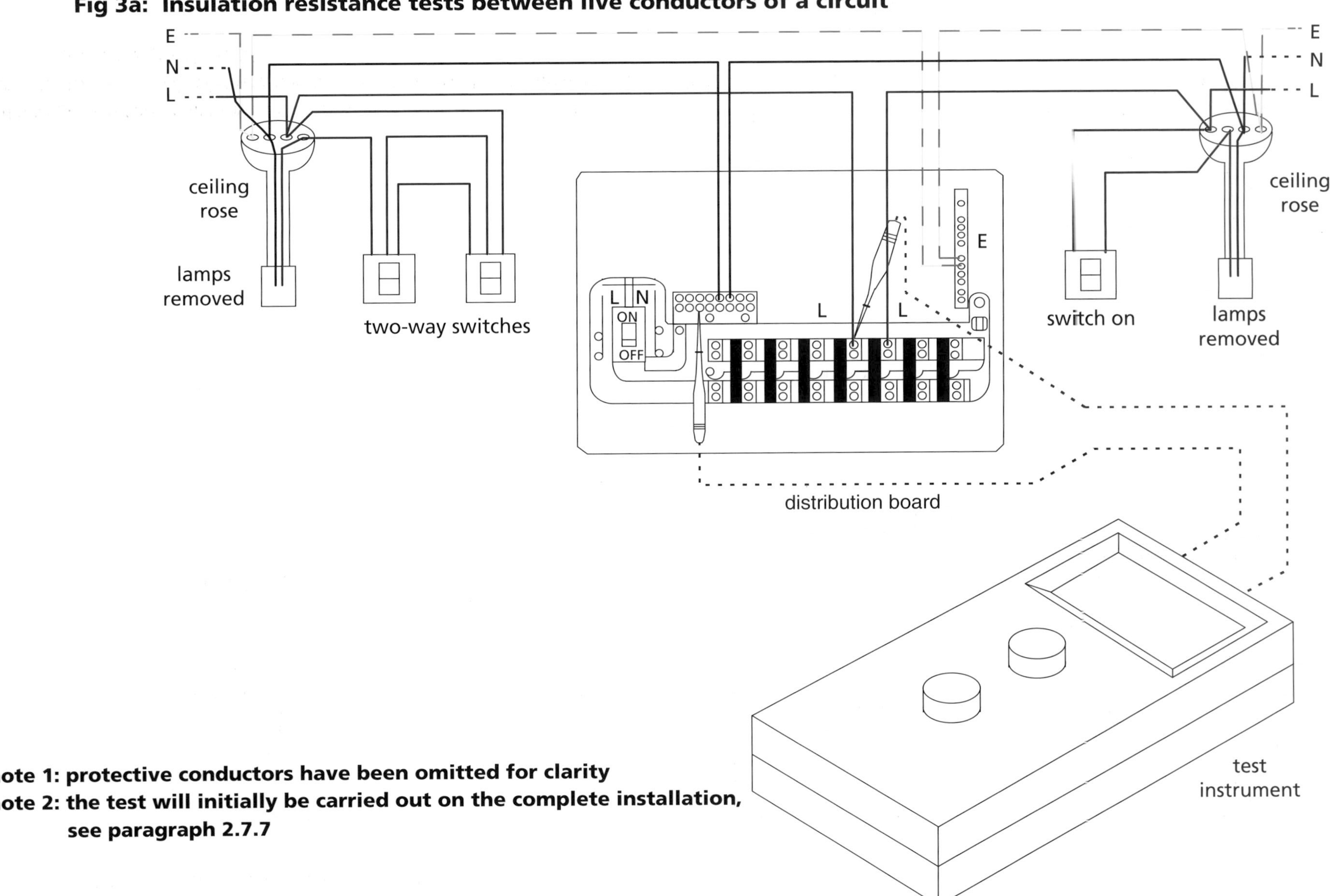

note 1: protective conductors have been omitted for clarity

note 2: the test will initially be carried out on the complete installation, see paragraph 2.7.7

Fig 3b: Insulation resistance tests to earth

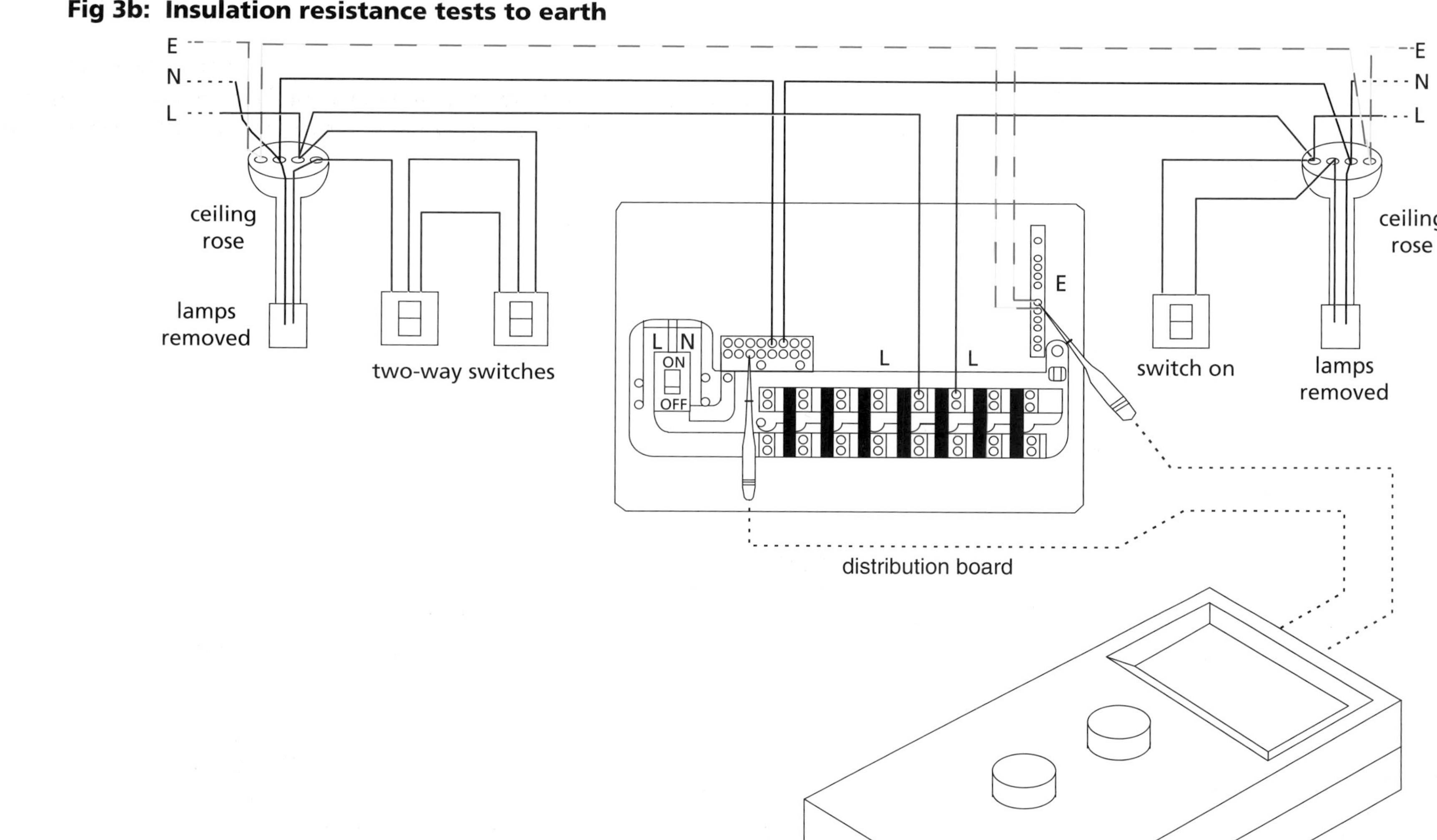

note 1: protective conductors have been omitted for clarity
note 2: the test will initially be carried out on the complete installation, see paragraph 2.7.7.

2.7.8 Site applied insulation

Part 3
412
713

Site applied insulation tests are carried out only where insulation is applied during erection. They are not applied when type-tested switchgear is assembled on site. The tests involve the use of high voltages and great care is necessary to avoid danger. 412-02 413-03

When protection against direct contact is afforded by insulation which has been applied to live parts of equipment during erection on site, a test should be made to check that the insulation is capable of withstanding an applied test voltage equivalent to that specified in the British Standard for similar factory-built equipment. 713-05-01

The test voltage is applied between the live conductors connected together, and metallic foil wrapped closely around all external surfaces of the insulation. The test voltage and duration must accord with the appropriate British Standard specification.

Where there is no such British Standard, the test may be applied using a test voltage of 3750 V a.c., rms. This test voltage should be at supply frequency, and should be applied to the insulation for a duration of 1 minute. The insulation can be deemed to be satisfactory if no breakdown or flashover occurs during the period of test.

Instrument: Use an applied voltage tester for this test. Refer to Section 4.5

Where protection against indirect contact is provided by supplementary insulation applied to equipment during erection, a test should be made to verify that: 713-05-02

(i) the insulating enclosure affords a degree of protection not less than IP2X or IPXXB. For details of test methods refer to Section 2.7.10 (Protection by barriers or enclosures)

(ii) the insulating enclosure is capable of withstanding, without breakdown or flashover, an applied test voltage equivalent to that specified in the British Standard for similar factory-built equipment.

Instrument: Use an applied voltage tester for this test. Refer to Section 4.5

2.7.9 Protection by separation of circuits

SELV

713-06-01
713-06-02

The source of the SELV supply should be inspected for conformity with Regulation 411-02-02 and if necessary the voltage should be measured to confirm that it does not exceed 50 V a.c. or 120 V d.c. If the voltage exceeds 25 V a.c. or 70 V d.c. (60 V d.c. ripple-free), the means of protection against direct contact should be verified. This may be by either: 411-02-02

(i) inspection of the barriers or enclosures to ensure they offer a degree of protection not less than IP2X (IPXXB) or IP4X as required by Regulation 412-03, or

(ii) inspection of the insulation. 411-02-09

The live conductors of any adjacent higher voltage circuit in contact with or in the same enclosure as SELV (separated extra-low voltage) circuits must be tested to verify electrical separation in accordance with BS 7671. Table 71A 411-02-05

This is achieved by testing between the live conductors of each SELV circuit connected together and the live conductors of any adjacent higher voltage circuits connected together.

Testing of SELV circuits

The first test applied to this arrangement is an insulation resistance test in accordance with Table 2.2. Table 71A

Where the circuit is supplied from a safety source complying with BS 3535, the test is made at 250 V d.c. and the required minimum insulation resistance (of Table 2.2) is 0.25 megohm. In practice readings less than 5 megohms require investigation. 411-02-02 411-02-04

Instrument: Use an insulation resistance tester for this test. Refer to Section 4.4.

The SELV circuit conductors should be inspected to verify compliance with BS 7671. 411-02-06

Tests between SELV circuits and other circuits

713-06-02 To demonstrate compliance with BS 7671 an additional insulation resistance test should be applied at 500 V d.c. between the live conductors of the SELV circuit and the protective conductor of the higher voltage circuit present. The insulation resistance must not be less than 0.5 megohm. In practice a reading less than 5 megohms requires investigation.

Instrument: Use an insulation resistance tester for this test. Refer to Section 4.4.

Compliance with Regulation 411-02-08 will have been verified by the inspection required by Regulation 411-02-06(iii). This should be confirmed. 411-02-08 411-02-06

The exposed-conductive-parts should be inspected to confirm compliance with Regulation 411-02-07. 411-02-07

PELV

713-06-01 PELV (protective extra-low voltage) installations are inspected and tested as for SELV installations except that an insulation test is not made between PELV circuits and earth. 471-14

713-06-03 (PELV systems may include a protective conductor connected to the protective conductor of the primary circuit. Protection against indirect contact in the secondary circuit is dependent upon the primary circuit protection).

Functional extra-low voltage

713-06-05 Extra-low voltage circuits not meeting the requirements for SELV or PELV are inspected and tested as low voltage circuits. 471-14-03

Electrical separation

713-06-04 The source of supply should be inspected to confirm compliance with the Regulations. In addition, should any doubt exist, the voltage should be measured to verify it does not exceed 500 V. 713-06-01 413-06-02(iv)

Where the source of supply does not comply with Regulation 413-06-02(i), compliance with Regulation 413-03 must be verified. Where this cannot be done or doubt exists, the insulation between live parts of the separated circuit and any other adjacent (in the same enclosure or touching) conductor, and/or to earth, must be tested. This test should be performed at a voltage of 500 V d.c. and the insulation resistance should be not less than 0.5 megohm. 413-06-02 (iii)(b) 413-03 Table 71A

Instrument: Use an insulation resistance tester for this test. Refer to Section 4.4.

The live parts of the separated circuit must be tested to ensure that they are electrically separate from other circuits. This is achieved by testing between the live conductors of the separated circuit connected together and the conductors of any other adjacent circuit strapped together. 413-06-02 (iii)(iv)

The first test applied to this arrangement is an insulation resistance test at 500 V d.c. The insulation resistance should be not less than 0.5 megohm.

Instrument: Use an insulation resistance tester for this test. Refer to Section 4.4.

A separate wiring system is preferred for electrical separation. If multicore cables or insulated cables in insulated conduit are used, all cables must be insulated to the highest voltage present and each conductor must be protected against overcurrent. 413-06-03(iii)

A 500 V d.c. insulation resistance test is performed between the exposed-conductive-parts of any item of connected equipment, and the protective conductor or exposed-conductive-parts of any other circuit, to confirm compliance with the Regulations. The insulation resistance should be not less than 0.5 megohm. 413-06-03(iv)

Instrument: Use an insulation resistance tester for this test. Refer to Section 4.4.

If the separated circuit supplies more than one item of equipment, the requirements of the Regulations shall be verified as follows: 413-06-05

(i) apply a continuity test between all exposed-conductive-parts of the separated circuit to ensure they are bonded together. This equipotential bonding is then subjected to a 500 V d.c. insulation resistance test between the protective conductor or exposed-conductive-parts of other circuits, or to extraneous-conductive-parts. The insulation resistance should be not less than 0.5 megohm.

Instrument: Use an insulation resistance tester for this test. Refer to Section 4.4.

(ii) all socket-outlets must be inspected to ensure that the protective conductor contact is connected to the equipotential bonding conductor

(iii) all flexible cables other than those feeding Class II equipment must be inspected to ensure that they embody a protective conductor for use as an equipotential bonding conductor

(iv) operation of the protective device must be verified by measurement of the fault loop impedances to the various pieces of connected equipment, with reference to the type and rating of the protective device for the separated circuit. If protection is provided by overcurrent devices, the appropriate value of loop impedance given by Regulation 413-02-08 (Tables 41B, 41C and 41D for 230 V systems) shall be determined. Table 41A Table 41B

Instrument: Use a loop impedance tester for this test. Refer to Section 4.4.

2.7.10 Protection by barriers or enclosures

Protection by barriers or enclosures provided during erection

713 This test is not applicable to barriers or enclosures of factory-built 713-07
412 equipment, but only to those provided on site during the course of 412-03-01
assembly or erection and therefore is seldom necessary. Where, 412-03-02
during erection, an enclosure or barrier is provided to afford 412-03-04
protection from direct contact, a degree of protection not less than IP2X or IPXXB is required. Readily accessible horizontal top surfaces shall have a degree of protection of at least IP4X.

The degree of protection afforded by IP2X is defined in BS EN 60529 412-03-01
as protection against the entry of 'Fingers or similar objects not exceeding 80 mm in length. Solid objects exceeding 12 mm in diameter'. The test is made with a metallic standard test finger (test finger 1 to BS 3042).

Both joints of the finger may be bent through 90 ° with respect to the axis of the finger, but in one and the same direction only. The finger is pushed without undue force (not more than 10 N) against any openings in the enclosure and, if it enters, it is placed in every possible position.

A SELV supply, not exceeding 50 V, in series with a suitable lamp is connected between the test finger and the live parts inside the enclosure. Conducting parts covered only with varnish or paint or protected by oxidation or by a similar process, shall be covered with a metal foil electrically connected to those parts which are normally live in service.

The protection is satisfactory if the lamp does not light.

The degree of protection afforded by IP4X is defined in BS EN 60529 412-03-02
as protection against the entry of 'Wires or strips of thickness greater than 1.0 mm, and solid objects exceeding 1.0 mm in diameter'.

The test is made with a straight rigid steel wire of 1 mm +0.05/0 mm diameter applied with a force of 1 N ±10 per cent. The end of the wire shall be free from burrs, and at a right angle to its length.

The protection is satisfactory if the wire cannot enter the enclosure.

Reference should be made to the appropriate product standard or BS EN 60529 for a fuller description of the degrees of protection, details of the standard test finger and other aspects of the tests.

2.7.11 Protection by a non-conducting location

Insulation of non-conducting floors and walls

413
713
Where protection against indirect contact is provided by a non-conducting location the following should be verified: 413-04 471-10 713-08

(i) exposed-conductive-parts should be arranged so that under normal circumstances a person will not come into simultaneous contact with: 413-04-02

two exposed-conductive-parts or

an exposed-conductive-part and any extraneous-conductive-part

(ii) in a non-conducting location there must be no protective conductors 413-04-03

(iii) any socket-outlets installed in a non-conducting location must not incorporate an earthing contact

(iv) the resistance of insulating floors and walls to the main protective conductor of the installation should be tested at not less than three points on each relevant surface, one of which should be not less than 1 m and not more than 1.2 m from any extraneous-conductive-part, e.g. pipes, in the location. Methods of measuring the insulation resistance/impedance of floors and walls are described later in this section. 413-04-04 713-08-01

If at any point the resistance is less than the specified value (50 kΩ where the voltage to earth does not exceed 500 V) the floors and walls are deemed to be extraneous-conductive-parts. 413-04-04

If any extraneous-conductive-part is insulated it should be tested and must be capable of withstanding at least 2 kV a.c. rms without breakdown, and should not pass a leakage current exceeding 1 mA in normal use. 413-04-07 (iii) 713-08-02

The following test involves the use of high voltage and adequate precautions should be taken to prevent danger. Consult HS(G)13.

This testing is done by applying metallic foil closely around all the insulation, and then firstly applying 2 kV a.c. rms between this foil and the main protective conductor for a duration of 1 minute. The insulation can be deemed to be satisfactory if no breakdown or flashover occurs during the test. This test is then followed by an insulation resistance test at 500 V d.c. across the same connections. The insulation resistance should be not less than 0.5 megohm.

Instrument: Use a high output applied voltage tester for the high voltage test. Refer to Section 4.5 for additional precautions.

Measuring insulation resistance of floors and walls

A magneto-ohmmeter or battery-powered insulation tester providing a no-load voltage of approximately 500 V (or 1000 V if the rated voltage of the installation exceeds 500 V) is used as a d.c. source.

The resistance is measured between the test electrode and the main protective conductor of the installation.

The test electrodes may be either of the following types. In case of dispute, the use of test electrode 1 is the reference method.

It is recommended that the test be made before the application of the surface treatment (varnishes, paints and similar products).

Test electrode 1

The electrode (see Fig 4a) is a 250 mm square metallic plate with a 270 mm square of damped water absorbent paper or cloth from which surplus water has been removed, placed between the metal plate and the surface being tested.

During the measurement a force of approximately 750 N (12 stone or 75 kg in weight) in the case of floors or 250 N in the case of walls is applied on the plate.

Fig 4a: Test electrode 1

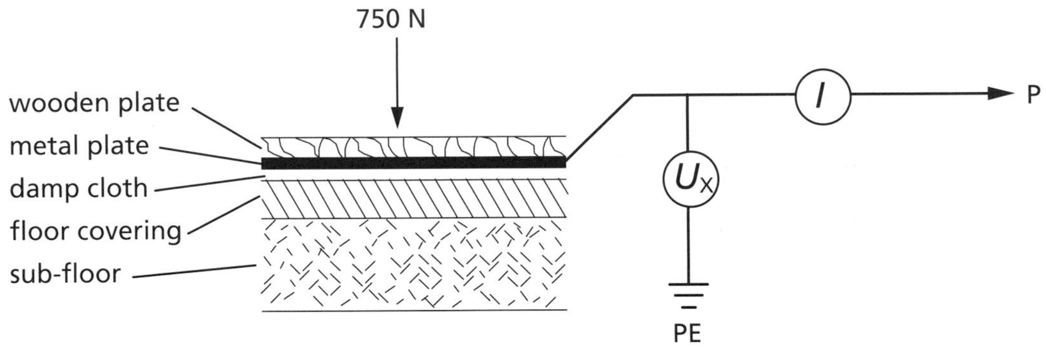

Test electrode 2

The test electrode (see Fig 4b) comprises a metallic tripod of which the parts resting on the floor form the points of an equilateral triangle. Each supporting part is provided with a flexible base ensuring, when loaded, close contact with the surface being tested over an area of approximately 900 mm^2 and having a combined resistance of less than 5000 Ω between the terminal and the conductive rubber pads.

Before measurements are made, the surface being tested is moistened or covered with a damp cloth. While measurements of the floors and walls are being made a force of approximately 750 N (floors) or 250 N (walls), is applied to the tripod.

Fig 4b: Test electrode 2

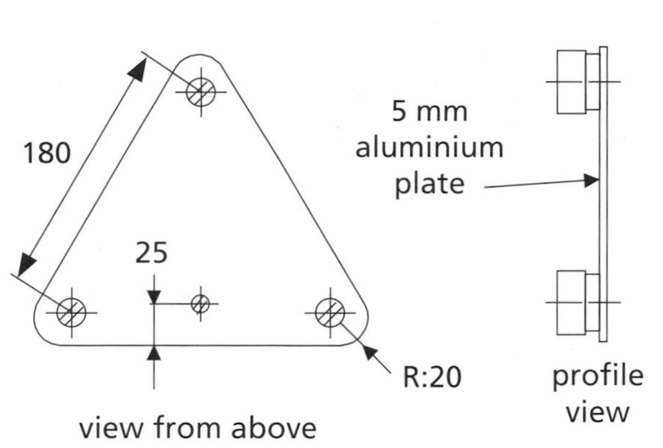

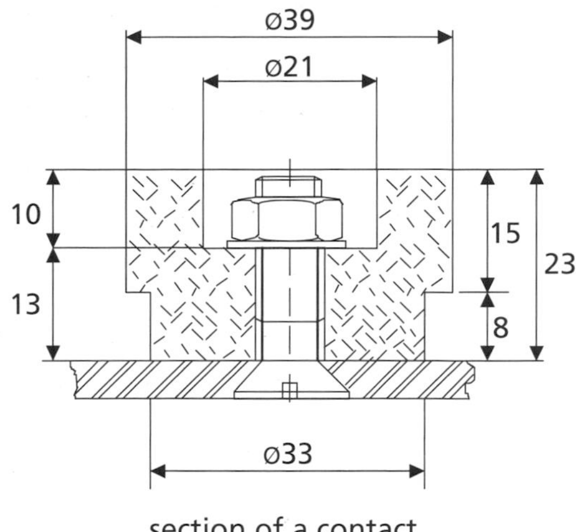

section of a contact
stud in conductive rubber

attached by screw,
washer and nut

Terminal

Contact stud in
conductive rubber

view from below

2.7.12 Polarity

713 The polarity of all circuits must be verified before connection to the supply, with either an ohmmeter or the continuity range of an insulation and continuity tester. 713-09

Instrument: Use a low-resistance ohmmeter for these tests. Refer to Section 4.3

One method of test is as described for Test method 1 paragraph 2.7.5 for testing the continuity of protective conductors. 713-02

For radial circuits the (R_1 + R_2) measurements, made as in Test method 1 paragraph 2.7.5, should be made at each point. 713-02

For ring circuits, if the test required by Regulation 713-03 has been carried out (paragraph 2.7.6), the correct connections of phase, neutral and circuit protective conductors will have been verified, subject to the following where applicable. 713-03

Different makes and types of accessories are not consistent in the relative position of the cable terminations to the socket tubes. Therefore, if the testing has been to the terminals on the reverse of the accessories and not to the socket tubes, a visual inspection is required.

It is necessary to check that all fuses and single-pole control and protective devices are connected in the phase conductor. The centre contact of screw type lampholders must be connected to the phase conductor (except E14 & E27 to BS EN 60238). The phase connection in socket-outlets and similar accessories is connected to the phase conductor.

See Fig 5 for Test method connections.

After connection of the supply, polarity should be confirmed using an approved voltage indicator.

Fig 5: Polarity test on a lighting circuit

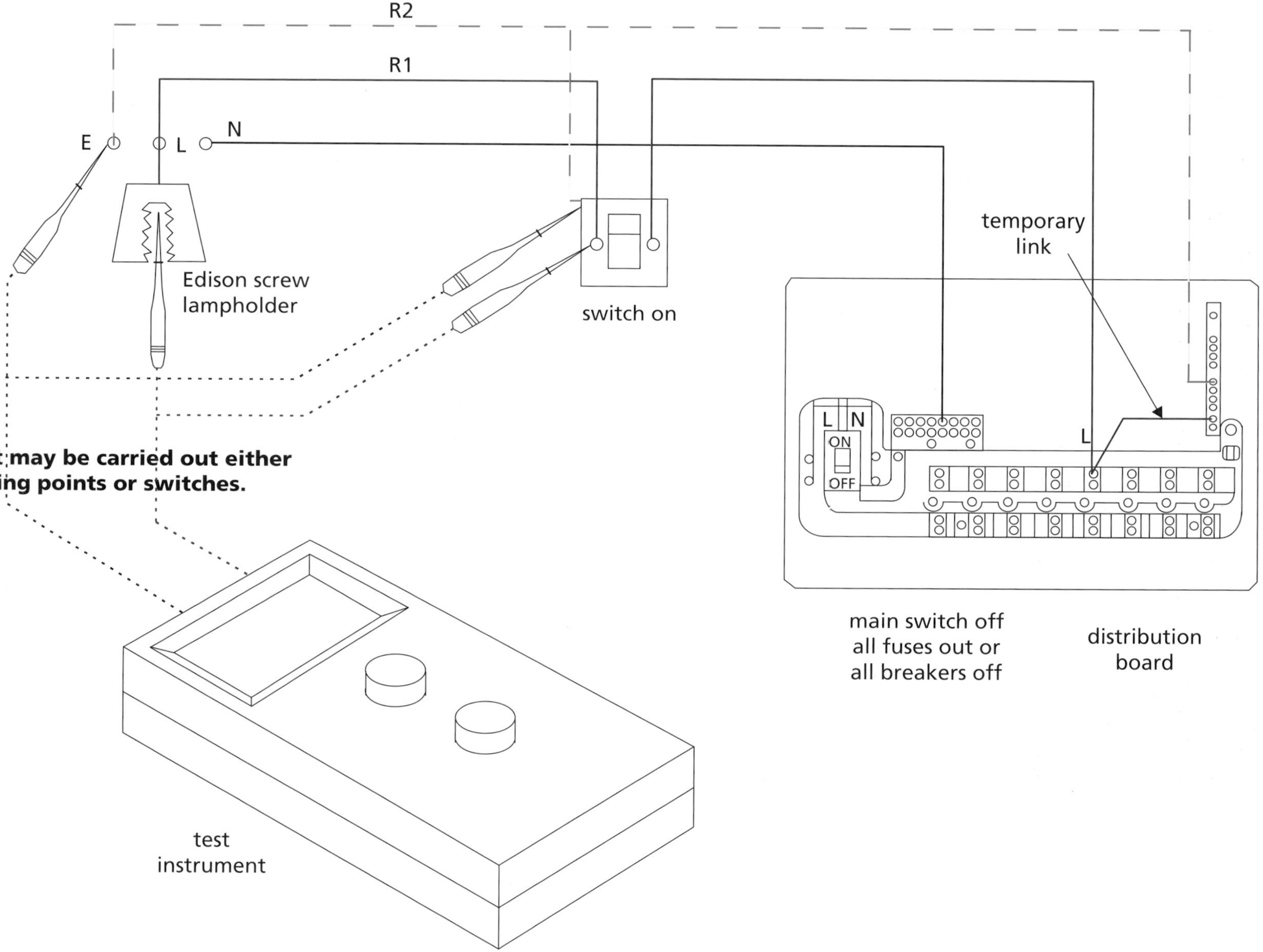

R2

R1

E

L

N

Edison screw lampholder

switch on

temporary link

L

L

N

ON

OFF

main switch off all fuses out or all breakers off

distribution board

note: The test may be carried out either at lighting points or switches.

test instrument

2.7.13 Earth electrode resistance

Measurement by standard method

713 When measuring earth electrode resistances to earth where low 713-10
413 values are required, as in the earthing of the neutral point of a 542-01
transformer or generator, test method 1 below may be used. 542-02

Instrument: Use an earth electrode resistance tester for this test. Refer to Section 4.7.

Test method 1

Before this test is undertaken, the earthing conductor to the earth electrode must be disconnected either at the electrode or at the main earthing terminal to ensure that all the test current passes through the earth electrode alone. This will leave the installation unprotected against earth faults. SWITCH OFF SUPPLY BEFORE DISCONNECTING THE EARTH.

The test should be carried out when the ground conditions are least favourable, such as during dry weather. 542-02-02

The test requires the use of two temporary test spikes (electrodes), and is carried out in the following manner.

Connection to the earth electrode is made using terminals C_1 and P_1 of a four-terminal earth tester. To exclude the resistance of the test leads from the resistance reading, individual leads should be taken from these terminals and connected separately to the electrode. If the test lead resistance is insignificant, the two terminals may be short-circuited at the tester and connection made with a single test lead, the same being true if using a three-terminal tester. Connection to the temporary spikes is made as shown in Fig 6.

The distance between the test spikes is important. If they are too close together, their resistance areas will overlap. In general, reliable results may be expected if the distance between the electrode under test and the current spike is at least ten times the maximum dimension of the electrode system, e.g. 30 m for a 3 m long rod electrode.

Three readings are taken: with the potential spike initially midway between the electrode and current spike, secondly at a position 10 per cent of the electrode-to-current spike distance back towards the electrode, and finally at a position 10 per cent of the distance towards the current spike.

By comparing the three readings, a percentage deviation can be determined. This is calculated by taking the average of the three readings, finding the maximum deviation of the readings from this average in ohms, and expressing this as a percentage of the average.

The accuracy of the measurement using this technique is typically 1.2 times the percentage deviation of the readings. It is difficult to achieve a measurement accuracy better than 2 per cent, and inadvisable to accept readings that differ by more than 5 per cent. To improve the accuracy of the measurement to acceptable levels, the test must be repeated with a larger separation between the electrode and the current spike.

Fig 6: Earth Electrode Test

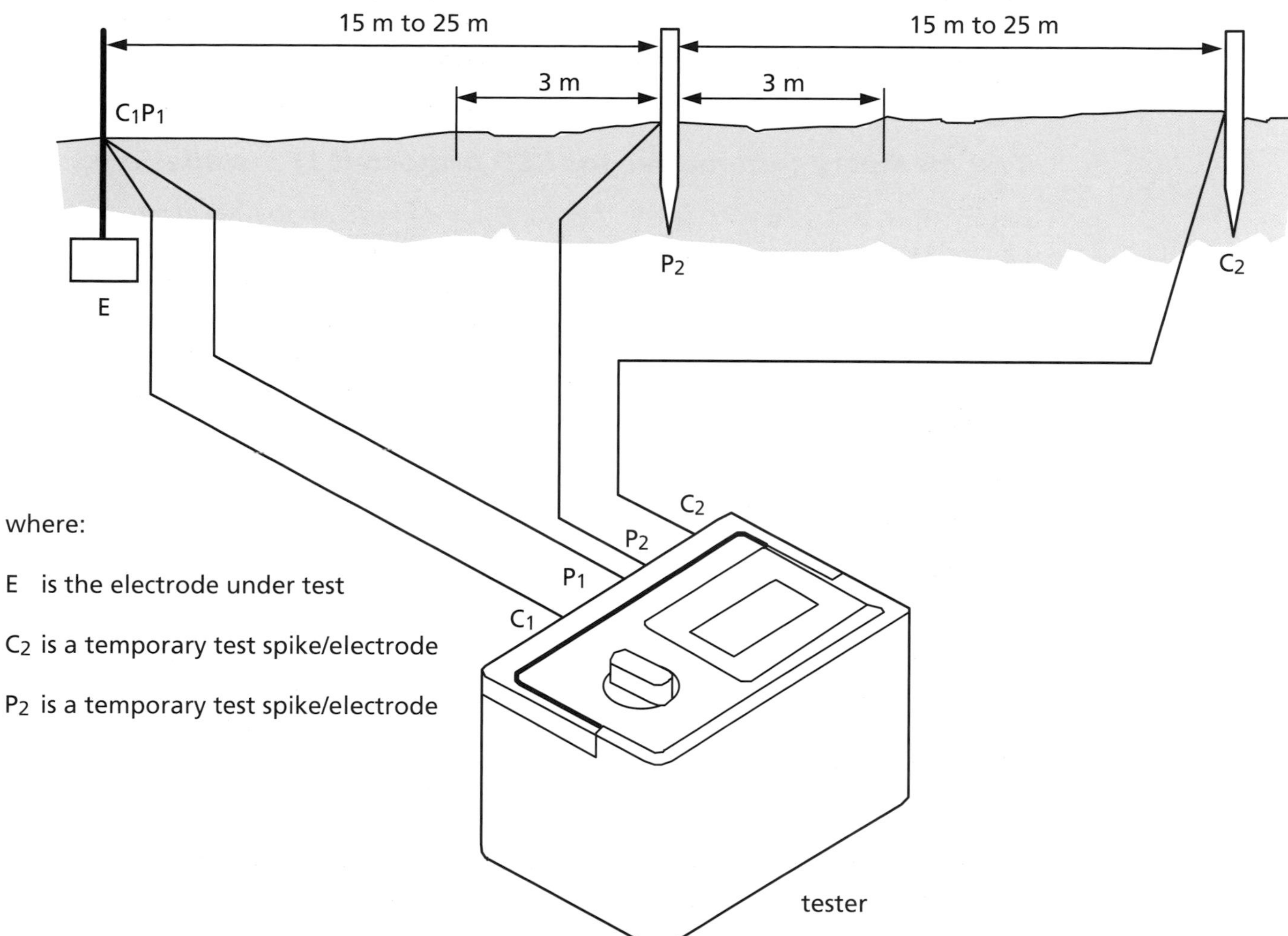

where:

E is the electrode under test

C_2 is a temporary test spike/electrode

P_2 is a temporary test spike/electrode

The instrument output current may be a.c. or reversed d.c. to overcome electrolytic effects. Because these testers employ phase-sensitive detectors (psd), the errors associated with stray currents are eliminated.

The instrument should be capable of checking that the resistance of the temporary spikes used for testing are within the accuracy limits stated in the instrument specification. This may be achieved by an indicator provided on the instrument, or the instrument should have a sufficiently high upper range to enable a discrete test to be performed on the spikes.

If the temporary spike resistance is too high, measures to reduce the resistance will be necessary, such as driving the spikes deeper into the ground or watering with brine to improve contact resistance. In no circumstances should these techniques be used to temporarily reduce the resistance of the earth electrode under test.

AFTER COMPLETION OF THE TESTING ENSURE THAT THE EARTHING CONDUCTOR IS RECONNECTED.

Earth electrode for RCDs

If the electrode under test is being used in conjunction with a residual current device the following method of test may be applied as an alternative to the earth electrode resistance test described above. In these circumstances, where the electrical resistances to earth are relatively high and precision is not required, an earth fault loop impedance tester may be used. Refer to Section 4.6.

Test method 2 (alternative for RCD protected TT installations)

Before this test is undertaken, the earthing conductor to the earth electrode should be disconnected at the main earthing terminal to ensure that all the test current passes through the earth electrode alone. This will leave the installation unprotected against earth faults. SWITCH OFF SUPPLY BEFORE DISCONNECTING THE EARTH.

A loop impedance tester is connected between the phase conductor at the source of the TT installation and the earth electrode, and a test performed. The impedance reading taken is treated as the electrode resistance.

BS 7671 requires:

$R_A I_{\Delta n} \leq 50$ V for normal dry locations 413-02-20

and $R_A I_{\Delta n} \leq 25$ for construction sites and agricultural premises 604-05-01 605-06-01

Where:

R_A is the sum of the resistances of the earth electrode and the protective conductor(s) connecting it to the exposed-conductive-part

$I_{\Delta n}$ is the rated residual operating current.

Maximum values of R_A for the basic standard ratings of residual current devices are given in the following table, unless the manufacturer declares alternative values.

TABLE 2.3 Maximum values of earth electrode resistance for TT installations

RCD rated residual operating current $I_{\Delta n}$	Maximum value of earth electrode resistance, R_A	
	normal dry locations	construction sites, agricultural and horticultural premises
30 mA	1660 Ω	830 Ω
100 mA	500 Ω	250 Ω
300 mA	160 Ω	80 Ω
500 mA	100 Ω	50 Ω

The table indicates that the use of a suitably rated RCD will theoretically allow much higher values of R_A, and therefore of Z_s, than could be expected by using the overcurrent devices for indirect contact shock protection. In practice, however, values above 200 ohms will require further investigation.

AFTER THE TEST ENSURE THAT THE EARTHING CONDUCTOR IS RECONNECTED.

Where items of stationary equipment having a protective conductor current exceeding 3.5 mA in normal service are supplied from an installation forming part of a TT system, the product of the total protective conductor current (in amperes) and twice the resistance of the installation earth electrode(s) (in ohms) must not exceed 50. 607-05-01

2.7.14 Earth fault loop impedance

Determining the earth fault loop impedance, Z_s

The earth fault current loop comprises the following elements, starting at the point of fault on the phase to earth loop:

— the circuit protective conductor

— the main earthing terminal and earthing conductor

— for TN systems the metallic return path or, in the case of TT and IT systems, the earth return path

— the path through the earthed neutral point of the transformer

— the source phase winding and

— the phase conductor from the source to the point of fault.

Earth fault loop impedance Z_s may be determined by: 713-11 413-02 Table 41B Table 41D Table 604B Table 605B

1) direct measurement using an earth loop impedance measurement instrument, see below, or

2) measurement of $R_1 + R_2$ during continuity testing (see 2.7.5) and adding to Z_e, i.e.

$$Z_s = Z_e + (R_1 + R_2)$$

Z_e is determined by:

— measurement - see below, or

— enquiry of the electricity supplier, or

— calculation.

Determining Z_e

Measurement

The reasons that Z_e is required to be measured are twofold:

1) to verify that there is an earth connection

2) to verify that the Z_e value is equal to or less than the value determined by the designer and used in the design calculations.

Z_e is measured using a phase-earth fault loop impedance tester at the origin of the installation.

The impedance measurement is made between the phase of the supply and the means of earthing with the main switch open or with all the circuits isolated. The means of earthing must be disconnected from the installation earthed equipotential bonding for the duration of the test to remove parallel paths. Care should be taken to avoid any shock hazard to the testing personnel and other persons on the site both whilst establishing contact, and performing the test. ENSURE THAT THE EARTH CONNECTION HAS BEEN REPLACED BEFORE RECLOSING THE MAIN SWITCH. 542-04-02 711-01-01

See Fig 7 for Test method connections.

Instrument: Use a loop impedance tester for this test. Refer to Part 4.

Measurement of (R_1 + R_2) to add to Z_e

Whilst testing the continuity of protective conductors of radial circuits, or whilst testing the continuity of ring final circuits, the value of $(R_1 + R_2)_{test}$ is measured (at the test ambient temperature).

The measured value of $(R_1 + R_2)_{test}$ for the final circuit should be added to the value of $(R_1 + R_2)_{test}$ for any distribution circuit supplying the final circuit, to give the total $(R_1 + R_2)_{test}$ from the origin of the installation.

Direct measurement of Z_s

Direct measurement of Z_s can only be made on a live installation. Neither the connection with earth nor bonding conductors are disconnected. Readings given by the loop impedance tester may be less than $Z_e + (R_1 + R_2)$ because of parallel earth return paths provided by any bonded extraneous-conductive-parts. This must be taken into account when comparing the results with design data.

Care should be taken during the tests to avoid any shock hazard to the testing personnel, other persons or livestock on site. 711-01-01

Fig 7: Test of Z_e at the origin of the installation

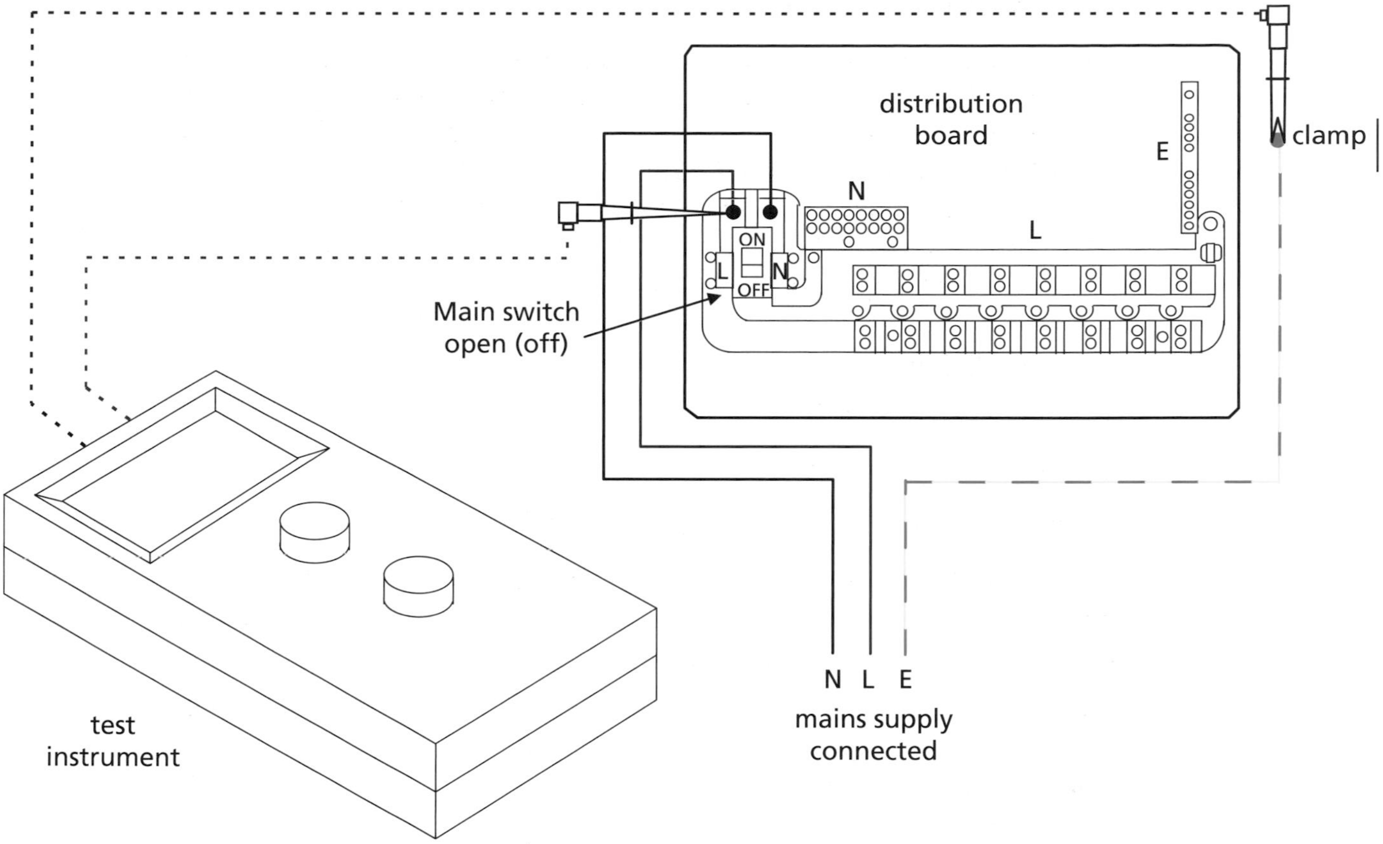

The diagram shows the test probes connected to phase and earth only. Three-wire loop impedance testers require a connection to the neutral for the instrument to operate.

Enquiry

Where Z_e is determined by enquiry of the electricity supplier, Z_s is then determined by adding (R_1 + R_2) to this value of Z_e. However, a test must be made to ensure that the electricity supplier's earth terminal is actually connected with earth, using a loop impedance tester or a test lamp.

Verification of test results

Values of Z_s should be compared with one of the following: 713-01-01

1) for standard thermoplastic (pvc) circuits, the values in Appendix 2 of this Guidance Note
2) earth fault loop impedance figures provided by the designer
3) tabulated values in BS 7671, corrected for temperature
4) rule of thumb figures.

Table 41B
Table 41D
Table 604B
Table 605B

1) Standard thermoplastic (pvc) circuits

The tabulated values given in Appendix 2 of this Guidance Note must not be exceeded when testing in an ambient temperature of 10 °C to 20 °C. As this is the normal temperature range to be expected, then correction for temperature is not usually required. Appendix 2 also provides a means of correcting these values for other test temperatures.

2) Earth fault loop impedance figures provided by the designer

This Guidance Note gives formulae which designers may use to obtain maximum values of Z_S for test purposes.

Examples of how these formulae may be used are given in Appendix 2.

3) Tabulated values in BS 7671, corrected for temperature

The note below the tables in BS 7671 states that if the conductors are tested at a temperature which is different to their maximum permitted operating temperature, which they usually will be, then the reading should be adjusted accordingly. Table 41B Table 41D Table 604B Table 605B

Appendix 2 provides a formula for making this adjustment together with a worked example.

4) Rule of Thumb figures

As a rule of thumb, the measured value of earth fault loop impedance for each circuit at the most remote outlet should not exceed three-quarters of the relevant value in the BS 7671 tables. The three-quarters figure allows for a reduced cross-section protective conductor and errs on the side of safety. Table 41B Table 41D Table 604B Table 605B

Alternative method

As an alternative to meeting the 0.4 second disconnection time of Regulation 413-02-08 (Tables 41B1 and 41B2) a disconnection time of 5 seconds is allowed for circuits supplying socket-outlets, portable equipment intended for manual movement during use, or hand-held Class I equipment, if R_2 does not exceed the values in Table 41C of BS 7671. 413-02-12 Table 41C

R_2 measured during the continuity tests of paragraph 2.7.5 is recorded in the Schedule of Test Results (col 7).

R_2 is corrected for temperature using Tables 1C or 1D in Appendix 1 of this Guidance Note and compared with the value in Table 41C of BS 7671.

Earth fault loop impedance test voltage

The normal method of test employed by a phase earth loop tester is to compare the unloaded loop circuit voltage with the circuit voltage when loaded with a low resistance, typically 10 ohms. This method of test can create an electric shock hazard if the phase earth loop impedance is high and the test duration is not limited. In these circumstances the potential of the protective conductor could approach phase voltage for the duration of the test (see paragraph 4.6).

Residual current devices

The test (measuring) current of earth fault loop impedance testers may trip any RCD protecting the circuit. This will prevent a measurement being taken and may result in an unwanted disconnection of supply to the circuit under test.

Instrument manufacturers can supply loop testers that are less liable to trip RCDs. There are two common techniques :

1) testing at limited current
2) d.c. biasing the RCD.

1) Limited current type

Some instruments limit the test current to below 15 mA. This should mean that RCDs with a rated residual operating current of 30 mA and greater will not trip.

2) d.c. biasing type

Loop testers using a d.c. biasing technique saturate the core of the RCD prior to testing so that the test current is not detected. This technique can usually be expected to be effective for both type A and type AC RCDs.

Note: Type A RCDs - tripping is ensured for residual sinusoidal alternating currents and for pulsating direct currents

Type AC RCDs - tripping is ensured for residual sinusoidal alternating currents.

2.7.15 Prospective fault current

Regulation 713-12-01 requires that the prospective fault current, I_{pf}, under both short-circuit and earth fault conditions, be measured, calculated or determined by another method, at the origin and at other relevant points in the installation. 713-12-01

Regulation 713-12-01 introduces the requirements of Regulation 434-02-01 into the testing section. The designer is required to determine the prospective fault current, under both short-circuit and earth fault conditions, **at every relevant point of the installation**. This may be done by calculation, ascertained by enquiry or by measurement. The expression 'every relevant point' means every point where a protective device is required to operate under fault conditions, and includes the origin of the installation. 434-02-01

Regulation 434-03-01 states that the breaking capacity rating of each protective device shall be not less than the prospective fault current at its point of installation. The term prospective fault current includes the prospective short-circuit current and the prospective earth fault current. It is the greater of these two prospective fault currents which should be determined and compared with the breaking capacity of the device. 434-03-01

Fig 8: Measurement of prospective short-circuit current

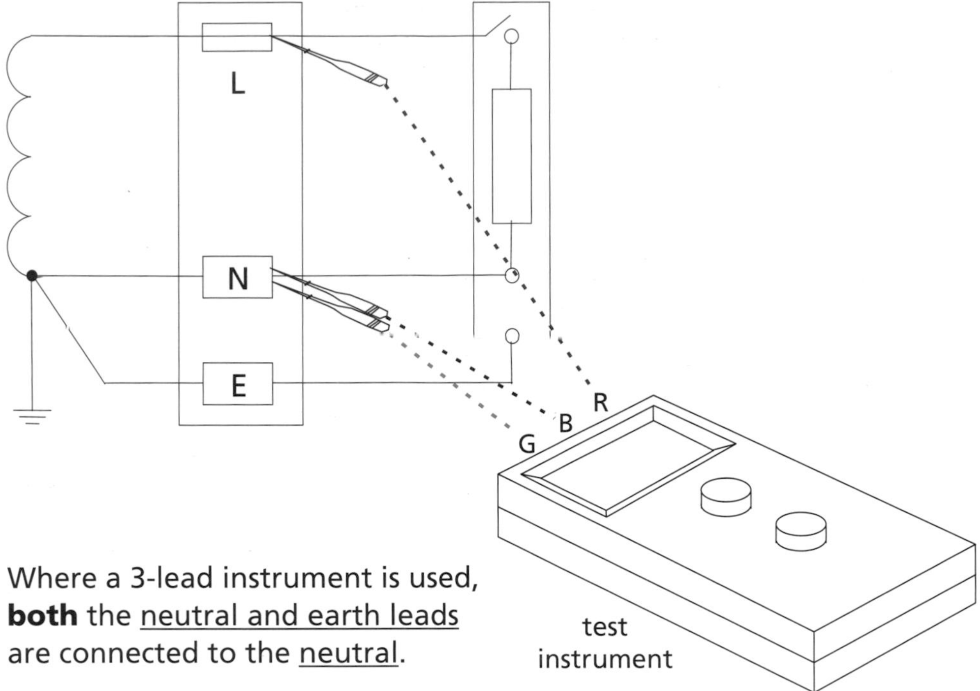

Where a 3-lead instrument is used, **both** the neutral and earth leads are connected to the neutral.

With the power on, the **maximum** value of the prospective short-circuit current can be obtained by direct measurement between live conductors at the protective device at the origin of the installation as shown in Fig 8. With many instruments, the voltage between phases can not be measured directly. For three-phase supplies, the maximum balanced prospective short-circuit level will be, as a rule of thumb, approximately twice the single-phase value. This figure errs on the side of safety.

These values then should be checked with the breaking capacity of the protective device to ensure that the breaking capacity is greater than the measured value of prospective fault

Fig 9: Measurement of prospective earth fault current

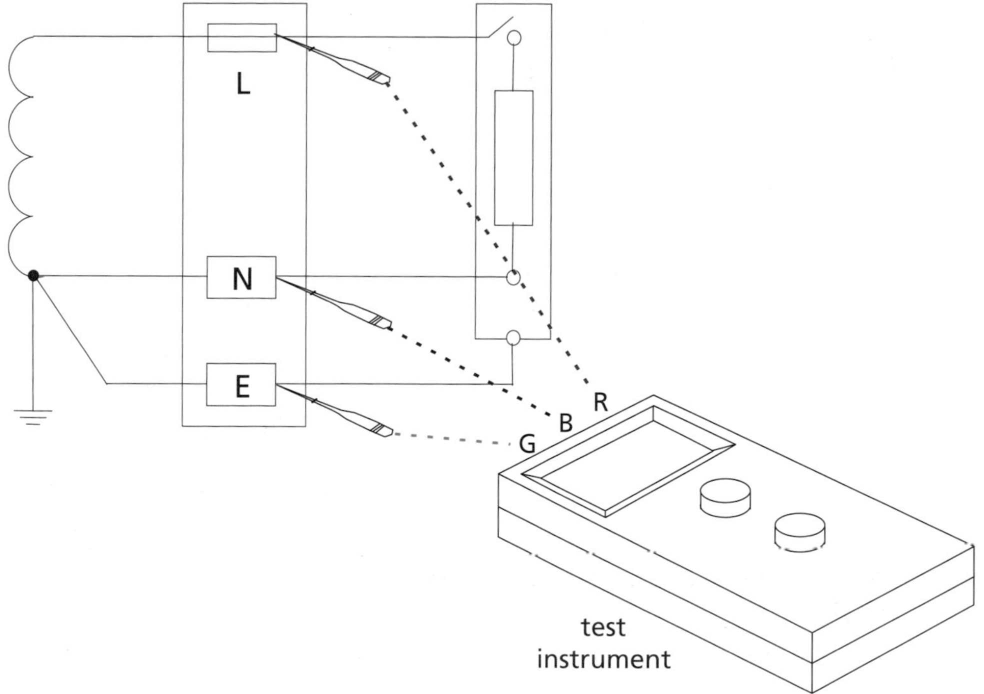

The prospective earth fault current may be measured with the same instrument, as indicated in Fig 9.

Instrument: Use a prospective fault current tester for this test. Refer to Section 4.6 - see final paragraph.

The larger of the two values (I_{pf}) should be recorded on the Schedule of Test Results. For a three-phase system, the prospective short-circuit current will always be larger than the earth fault current.

Rated short-circuit breaking capacities of protective devices

The rated short-circuit capacities of fuses, and circuit-breakers to BS 3871 (now withdrawn) and BS EN 60898 - are shown in Table 2.4.

BS 3871 identified the short-circuit capacity of circuit-breakers with an 'M' rating.

TABLE 2.4 Rated Short-circuit Capacities

Device type	Device designation	Rated short-circuit capacity kA
Semi-enclosed fuse to BS 3036 with category of duty	S1A S2A S4A	1 2 4
Cartridge fuse to BS 1361 type I type 2		16.5 33.0
General purpose fuse to BS 88-2 1		50 at 415 V
General purpose fuse to BS 88-6		16.5 at 240 V 80 at 415 V
Circuit-breakers to BS 3871 (replaced by BS EN 60898)	M1 M1.5 M3 M4.5 M6 M9	1 1.5 3 4.5 6 9
Circuit-breakers to BS EN 60898* and RCBOs to BS EN 61009*		I_{cn} I_{cs} 1.5 (1.5) 3.0 (3.0) 6 (6.0) 10 (7.5) 15 (7.5) 20 (10.0) 25 (12.5)

* Two rated short-circuit ratings are defined in BS EN 60898 and BS EN 61009

(a) I_{cn} the rated short-circuit capacity (marked on the device).

(b) I_{cs} the service short-circuit capacity.

The difference between the two is the condition of the circuit-breaker after manufacturer's testing.

I_{cn} is the maximum fault current the breaker can interrupt safely, although the breaker may no longer be usable.

I_{cs} is the maximum fault current the breaker can interrupt safely without loss of performance.

The I_{cn} value is marked on the device in a rectangle e.g. [6000] and for the majority of applications the prospective fault current at the terminals of the circuit-breaker should not exceed this value.

For domestic installations the prospective fault current is unlikely to exceed 6 kA up to which value the I_{cn}

Where a service cut-out containing a cartridge fuse to BS 1361 type 2 supplies a consumer unit which complies with BS 5486-13 or BS EN 60439-3, then the short-circuit capacity of the overcurrent protective devices within consumer units may be taken to be 16 kA.

Fault currents up to 16 kA

Except for London and some other major city centres, the maximum fault current for 230 V single-phase supplies up to 100 A will not exceed 16 kA.

The short-circuit capacity of overcurrent protective devices incorporated within consumer units may be taken to be 16 kA where:

- the current ratings of the devices do not exceed 50 A.
- the consumer unit complies with BS 5486-13 or BS EN 60439-3.
- The consumer unit is supplied through a type 2 fuse to BS 1361 : 1971 rated at no more than 100 A.

Recording the prospective fault current

Both the Electrical Installation Certificate and the Periodic Inspection Report contain a box headed Nature of Supply Parameters, which requires the prospective fault current at the origin to be recorded. The value to be recorded is the greater of either the short-circuit current (between live conductors) or the earth fault current (between phase conductor(s) and the main earthing terminal). If it is considered necessary to record values at other relevant points, they can be recorded on the Schedule of Test Results. Where the protective devices used at the origin have the necessary rated breaking capacity and similar devices are used throughout the installation, it can be assumed that the Regulations are satisfied in this respect for all distribution boards.

2.7.16 Functional testing

Operation of residual current devices

While the following tests are not a specific requirement of BS 7671, it is recommended that they are carried out.

713-13-01 In order to test the effectiveness of residual current devices after installation a series of tests may be applied to check that they are within specification and operate satisfactorily. This test sequence will be in addition to proving that the test button is operational. The effectiveness of the test button should be checked after the test sequence.

For each of the tests readings should be taken on both positive and negative half cycles and the longer operating time recorded.

Prior to these RCD tests it is essential, for safety reasons, that the earth loop impedance is tested to check the requirements have been met. 413-02-16

Instrument: Use an RCD tester for these tests. Refer to Section 4.8.

Test method

The test is made on the load side of the RCD between the phase conductor of the protected circuit and the associated cpc. The load should be disconnected during the test.

RCD testers require a few milliamperes to operate the instrument, and this is normally obtained from the phase and neutral of the circuit under test. When testing a three-phase RCD protecting a three-wire circuit, the instrument's neutral is required to be connected to earth. This means that the test current will be increased by the instrument supply current and will cause some devices to operate during the 50 per cent test at a time when they should not operate. Under this circumstance it is necessary to check the operating parameters of the RCD with the manufacturer before failing the device.

These tests can result in a potentially dangerous voltage on exposed- 711-01-01
and extraneous-conductive-parts when the earth fault loop impedance

approaches the maximum acceptable limits. Precautions must therefore be taken to prevent contact of persons or livestock with such parts.

General purpose RCDs to BS 4293

(i) with a leakage current flowing equivalent to 50 % of the rated tripping current of the RCD, the device should not open

(ii) with a leakage current flowing equivalent to 100 % of the rated tripping current of the RCD, the device should open in less than 200 ms.

Where the RCD incorporates an intentional time delay it should trip within a time range from 50 % of the rated time delay plus 200 ms to 100 % of the rated time delay plus 200 ms.

Because of the variability of the time delay it is not possible to specify a maximum test time. It is therefore imperative that the circuit protective conductor does not rise more than 50 V above earth potential ($Z_s I_{\Delta n} \leq 50$ V). It is suggested that in practice a 2 s 413-02-16
maximum test time is sufficient.

General purpose RCDs to BS EN 61008 or RCBOs to BS EN 61009

(i) with a leakage current flowing equivalent to 50 % of the rated tripping current of the RCD, the device should not open

(ii) with a leakage current flowing equivalent to 100 % of the rated tripping current of the RCD, the device should open in less than 300 ms unless it is of "Type S" (or selective) which incorporates an intentional time delay, when it should trip within the time range from 130 ms to 500 ms.

RCD protected socket-outlets to BS 7288

(i) with a test current flowing equivalent to 50 % of the rated tripping current of the RCD, the device should not open

(ii) with a test current flowing equivalent to 100 % of the rated tripping current of the RCD, the device should open in less than 200 ms.

Additional requirement for supplementary protection

Where an RCD with a rated residual operating current, $I_{\Delta n}$, not 412-06-02
exceeding 30 mA is used to provide supplementary protection against direct contact, the operating time of the device must not exceed 40 ms when subjected to a test current of 5 $I_{\Delta n}$. The maximum test time must not be longer than 40 ms, unless the protective conductor potential rises by less than 50 V. (The instrument supplier will advise on compliance).

Integral test device

An integral test device is incorporated in each RCD. This device 514-12-02
enables the mechanical parts of the RCD to be verified by pressing the button marked 'T' or 'Test'.

Operation of the integral test device does not provide a means of checking:

(i) the continuity of the earthing conductor or the associated circuit protective conductors, or

(ii) any earth electrode or other means of earthing, or

(iii) any other part of the associated installation earthing

(iv) the sensitivity of the device.

The RCD test button will only operate the RCD if it is energised.

Functional checks

All assemblies, including switchgear, controls and interlocks, are to be functionally tested, that is operated to confirm that they work and
713-12-02 are properly installed, mounted and adjusted.

Section 3 — Periodic Inspection and Testing

3.1 Purpose of periodic inspection and testing

731
732

The purpose of periodic inspection and testing is to provide, so far as is reasonably practicable, for: 731-01-04

(i) the safety of persons and livestock against the effects of electric shock and burns in accordance with the general requirements of Regulation 130-01, and

(ii) protection against damage to property by fire and heat arising from an installation defect, and

(iii) confirmation that the installation is not damaged or deteriorated so as to impair safety, and

(iv) the identification of installation defects and non-compliance with the requirements of the Regulations which may give rise to danger.

For an installation under effective supervision in normal use, periodic inspection and testing may be replaced by an adequate regime of continuous monitoring and maintenance of the installation and all its constituent equipment by skilled persons. Appropriate maintenance records must be kept. 732-01-02

If an installation is maintained under a planned maintenance management system, incorporating monitoring and supervised by a suitably qualified electrical engineer, with the results being recorded and kept over a period of time, then a formal periodic inspection and test certificate may not be required. The records may be kept on paper or computer and should record that electrical maintenance and testing has been carried out. The results of any tests should be recorded. The results should be available for scrutiny and need not be in the standard IEE Periodic Inspection Report format.

3.2 Necessity for periodic inspection and testing

731

Periodic inspection and testing is necessary because all electrical installations deteriorate due to a number of factors such as damage, wear, tear, corrosion, excessive electrical loading, ageing and environmental influences. Consequently:

(i) legislation requires that electrical installations are maintained in a safe condition and therefore must be periodically inspected and tested — Tables 3.1 and 3.2

(ii) licensing authorities, public bodies, insurance companies, mortgage lenders and others may require periodic inspection and testing of electrical installations — Tables 3.1 and 3.2

(iii) additionally, periodic inspection and testing should be considered:

(a) to assess compliance with BS 7671

(b) on a change of occupancy of the premises

(c) on a change of use of the premises

(d) after alterations or additions to the original installation

(e) because of any significant change in the electrical loading of the installation

(f) where there is reason to believe that damage may have been caused to the installation.

Reference to legislation and other documents is made below and it is vital that these requirements are ascertained before undertaking periodic inspection and testing.

3.3 Electricity at Work Regulations

Regulation 4(2) of the Electricity at Work Regulations 1989 requires that:

As may be necessary to prevent danger, all systems shall be maintained so as to prevent, so far as is reasonably practicable, such danger.

The Memorandum of Guidance published by the Health and Safety Executive advises that this regulation is concerned with the need for maintenance to ensure the safety of the system rather than being concerned with the activity of doing the maintenance in a safe manner, which is required by Regulation 4(3). The obligation to maintain a system arises if danger would otherwise result. There is no specific requirement to carry out a maintenance activity as such, what is required is that the system be kept in a safe condition. The frequency and nature of the maintenance must be such as to prevent danger so far as is reasonably practicable. Regular inspection of equipment including the electrical installation is an essential part of any preventive maintenance programme. This regular inspection may be carried out as required with or without dismantling and supplemented by testing.

There is no specific requirement to test the installation on every inspection. Where testing requires dismantling, the tester should consider whether that the risks associated with dismantling and re-assembling are justified. Dismantling, and particularly disconnection of cables or components, introduces a risk of unsatisfactory reassembly.

3.4 Design

When carrying out the design of an installation and particularly when specifying the equipment, the designer will be taking into account the quality of the maintenance to be specified including the frequency of routine checks and the period between subsequent inspections (supplemented as necessary by testing).

Information on the requirements for routine checks and inspections should be provided in accordance with Section 6 of the Health and Safety at Work etc Act 1974 and as required by The Construction (Design and Management) Regulations 1994. Users of a premises should seek this information as the basis on which to make their own assessments. The Health and Safety Executive advise in their Memorandum of Guidance on the Electricity at Work Regulations 1989, that practical experience of an installation's use may indicate the need for an adjustment to the frequency of checks and inspections. This is a matter of judgement for the duty holder. The Electrical Installation Certificate requires the designer's advice as to the intervals between inspections to be inserted on the certificate.

3.5 Routine checks

Electrical installations should not be left without any attention for the periods of years that are normally allowed between formal inspections. In domestic premises it is presumed that the occupier will soon notice any breakages or excessive wear and arrange for precautions to be taken and repairs to be carried out. In other situations, there must be arrangements made for initiating reports of wear and tear from users of the premises. This should be supplemented by routine checks. The frequency of these checks will depend entirely upon the nature of the premises. Routine checks would typically include:

TABLE 3.1
Routine Checks

Activity	Check
Defects Reports	All reported defects have been rectified
Inspection	Look for: breakages wear/deterioration signs of overheating missing parts (covers, screws) switchgear accessible (not obstructed) doors of enclosures secure adequate labelling loose fixings
Operation	Operate: switchgear (where reasonable) equipment - switch on and off including RCDs (using test button)

These routine checks need not be carried out by an electrically skilled person but should be by somebody who is able to safely use the installation and recognise defects.

3.6 Required information

It is essential that the inspector knows the extent of the installation to be inspected and any criteria regarding the limit of the inspection. This should be recorded.

Enquiries should be made to the person responsible for the electrical installation with regard to the provision of diagrams, design criteria, electricity supply and earthing arrangements. 514-09

Diagrams, charts or tables should be available to indicate the type and composition of circuits, identification of protective devices for shock protection, isolation and switching and a description of the method used for protection against indirect contact.

3.7 Frequency of inspection

The frequency of periodic inspection and testing must be determined taking into account:

(i) the type of installation 732-01-01

(ii) its use and operation

(iii) the frequency and quality of maintenance

(iv) the external influences to which it is subjected.

Table 3.2 provides guidance on the frequency of formal inspections of electrical installations as well as the routine checks. The 'initial frequencies' in the title of the table refers to the time interval between the issuing of the Electrical Installation Certificate on completion of the work and the first inspection. The recommended frequency of subsequent inspections may be increased or decreased at the discretion of the person carrying out the inspection and testing.

As well as for domestic and commercial premises, a change in occupancy of other premises may necessitate additional inspection and testing.

The formal inspections should be carried out in accordance with Chapter 73 of BS 7671.This requires an inspection comprising careful 731-01-03 scrutiny of the installation, carried out without dismantling or with partial dismantling as required, together with the appropriate tests of Chapter 71.

TABLE 3.2 Recommended Initial Frequencies of Inspection of Electrical Installations

Type of installation 1	Routine check sub-clause 3.5 2	Maximum period between inspections and testing as necessary 3	Reference (see notes below) 4
General installation			
Domestic	----	Change of occupancy/10 years	
Commercial	1 year	Change of occupancy/ 5 years	1, 2
Educational establishments	4 months	5 years	1, 2
Hospitals	1 year	5 years	1, 2
Industrial	1 year	3 years	1, 2
Residential accommodation	at change of occupancy/1 year	5 years	1
Offices	1 year	5 years	1, 2
Shops	1 year	5 years	1, 2
Laboratories	1 year	5 years	1, 2
Buildings open to the public			
Cinemas	1 year	3 years	2, 6, 7
Church installations	1 year	5 years	2
Leisure complexes (excluding swimming pools)	1 year	3 years	1,2,6
Places of public entertainment	1 year	3 years	1, 2, 6
Restaurants and hotels	1 year	5 years	1,2,6
Theatres	1 year	3 years	2, 6, 7
Public houses	1 year	5 years	1,2, 6
Village halls/Community centres	1 year	5 years	1, 2
Special installations			
Agricultural and horticultural	1 year	3 years	1, 2
Caravans	1 year	3 years	
Caravan Parks	6 months	1 year	1, 2, 6
Highway power supplies	as convenient	6 years	
Marinas	4 months	1 year	1, 2
Fish farms	4 months	1 year	1, 2
Swimming pools	4 months	1 year	1, 2, 6
Emergency lighting	Daily/monthly	3 years	2, 3, 4
Fire alarms	Daily/weekly/monthly	1 year	2, 4,5
Launderettes	1 year	1 year	1, 2, 6
Petrol filling stations	1 year	1 year	1, 2, 6
Construction site installations	3 months	3 months	1, 2

Reference Key

1. Particular attention must be taken to comply with SI 1988 No 1057. The Electricity Supply Regulations 1988 (as amended).
2. SI 1989 No 635. The Electricity at Work Regulations 1989 (Regulation 4 & Memorandum).
3. See BS 5266: Part 1: 1988 Code of practice for the emergency lighting of premises other than cinemas and certain other specified premises used for entertainment.
4. Other intervals are recommended for testing operation of batteries and generators.
5. See BS 5839: Part 1: 1988 Code of practice for system design installation and servicing (Fire detection and alarm systems for buildings).
6. Local Authority Conditions of Licence.
7. SI 1995 No 1129 (Clause 27) The Cinematograph (Safety) Regulations.

3.8 Requirements for inspection and testing

3.8.1 General procedure

Where diagrams, charts or tables are not available, a degree of 133
exploratory work may be necessary so that inspection and testing can 712-01-03(xvii)
be carried out safely and effectively. A survey may be necessary to 514-09
identify switchgear, controlgear, and the circuits they control.

Note should be made of any known changes in environmental conditions, building structure, and alterations or additions which have affected the suitability of the wiring for its present load and method of installation.

During the inspection, the opportunity should be taken to identify 711
dangers which might arise during the testing. Any location and 731-01-05
equipment for which safety precautions may be necessary should be noted and the appropriate steps taken.

Periodic tests should be made in such a way as to minimise disturbance of the installation and inconvenience to the user. Where it is necessary to disconnect part or the whole of an installation in order to carry out a test, the disconnection should be made at a time agreed with the user and for the minimum period needed to carry out the test. Where more than one test necessitates a disconnection where possible they should be made during one disconnection period.

A careful check should be made of the type of equipment on site so 713-04-02
that the necessary precautions can be taken, where conditions 713-04-04
require, to disconnect or short-out electronic and other equipment which may be damaged by testing. Special care must be taken where control and protective devices contain electronic components.

If the inspection and testing cannot be carried out safely without 514-09
diagrams or equivalent information, Section 6 of the Health and Safety at Work etc Act 1974 can be interpreted to require their preparation.

3.8.2 Scope

731 The requirement of BS 7671 for Periodic Inspection and Testing is for 731-01-03
INSPECTION comprising careful scrutiny of the installation without dismantling, or with partial dismantling as required, together with the tests of Chapter 71 considered appropriate by the person carrying out the inspection and testing. The scope of the periodic inspection and testing must be decided by a competent person, taking into account the availability of records and the use, condition and nature of the installation.

Consultation with the client or the client's representative prior to the periodic inspection and testing work being carried out is essential to determine the degree of disconnection which will be acceptable before planning the detailed inspection and testing.

For safety, it is necessary to carry out a visual inspection of the installation before testing or opening enclosures, removing covers, etc. So far as is reasonably practicable, the visual inspection must verify that the safety of persons, livestock and property is not endangered.

A thorough visual inspection should be made of all electrical equipment which is not concealed, and should include the accessible internal condition of a sample of the equipment. The external condition should be noted and if damage is identified or if the degree of protection has been impaired, this should be recorded on the schedule to the Report. The inspection should include a check on the condition of all electrical equipment and material, taking into account any available manufacturer's information, with regard to the following:

(i) safety
(ii) wear and tear
(iii) corrosion
(iv) damage
(v) excessive loading (overloading)
(vi) age
(vii) external influences
(viii) suitability.

The assessment of condition should take account of known changes in conditions influencing and affecting electrical safety, e.g. extraneous-conductive-parts, plumbing, structural changes.

Where sections of an electrical installation are excluded from the scope of a Periodic Inspection and Test, they should be identified in the "extent and limitations" box of the Report. However, such sections must not be permanently excluded from inspection and testing, and a suitable programme should be devised which includes the inspection and testing of such sections.

3.8.3 Isolation of supplies

The requirement of Regulation 14 of the Electricity at Work Regulations 1989 regarding working on or near live parts, must be observed during inspection of an installation.

In domestic type premises the whole installation can be readily isolated for inspection, but with most other installations it is not practicable and too disruptive to isolate the whole installation for the amount of time that is required for a comprehensive inspection. Much of the inspection in such premises has to be done whilst the installation is in operation.

Main switch panels can rarely be isolated from the supply for long periods; similarly, the disruption that may be caused by isolating final circuit distribution boards for long periods often cannot be tolerated.

Distribution boards should be isolated separately for short periods for the internal inspection of live parts and examination of connections.

Where it is necessary to inspect live parts inside equipment the supply to the equipment must be disconnected.

In order to minimise disruption to the operation of premises the appropriate supplementary testing in Section 3.10 should be applied at the same time as the inspection.

3.9 Periodic inspection

3.9.1 Comments on individual items to be inspected

731 a *Joints and connections*

It is not practicable to inspect every joint and termination in an electrical installation. Nevertheless a sample inspection should be made. An inspection should be made of all accessible parts of the electrical installation e.g. switchgear, distribution boards, and a sample of luminaire points and socket-outlets, to ensure that all terminal connections of the conductors are properly installed and secured. Any signs of overheating of conductors, terminations or equipment should be thoroughly investigated and included in the Report. 712-01-03(i) 712-01-03(vi)

b *Conductors*

The deterioration of, or damage to, conductors and their insulation, and their protective coverings, if any, should be noted.

The insulation and protective covering of each conductor at each distribution board of the electrical installation and at a sample of switchgear, luminaires, socket-outlets and other points, should be inspected to determine their condition and correct installation. There should be no signs of overheating, overloading or damage to the insulation, armour, sheath or conductors. 712-01-03(iv)

c *Flexible cables and cords*

Where a flexible cable or cord forms part of the fixed wiring installation, the inspection should include: 712-01-03(iii) 522

(i) examination of the cable or cord for damage or defects

(ii) examination of the terminations and anchorages for damage or defects

(iii) the correctness of its installation with regard to additional mechanical protection, heat resistant sleeving, etc.

d *Accessories and switchgear*

It is recommended that a random sample of a minimum of 10 per cent of all accessories and switchgear is given a thorough internal visual inspection of accessible parts to assess their electrical and mechanical condition. Where the inspection reveals: 712-01-03(v) 712-01-03(vi)

(i) results significantly different from results recorded previously

(ii) results significantly different from results reasonably to be expected

(iii) adverse conditions, e.g. fluid ingress or worn or damaged mechanisms,

the inspection should be extended to include every switching device associated with the installation under inspection unless there is clear evidence of how the damage occurred.

e *Protection against thermal effects*

The presence of fire barriers, seals and means of protection against thermal effects should be verified, if reasonably practicable. 712-01-03(vii)

f *Protection against direct and indirect contact* 712-01-03 (viii)(a)

SELV is commonly used as protection against direct and indirect contact. The requirements of 411-02 need to be checked, particularly 411
with respect to the source e.g. a safety isolating transformer to 411-02-02
BS 3535, the need to separate the circuits, and the segregation of 411-02-06
exposed-conductive-parts of the SELV system from any connection 411-02-07
with the earthing of the primary circuits, or from any other connection with earth.

g *Protection against direct contact*

It should be established that the means of protection against direct contact with live conductors is provided by one or more of the following methods: 712-01-03 (viii)(b) 412

(i) insulation of live parts 713-04

(ii) installation of barriers or enclosures 713-07

(iii) obstacles

(iv) placing out of reach

(v) SELV or PELV. 713-06

It should be established that the means of protection against direct contact with any live conductor meets the requirements for the safety of any person, livestock and property from the effects of electric shock, fire and burns.

For each method of protection against direct contact it should be established that there has been no deterioration or damage to insulation, no removal of barriers or obstacles and no alterations to enclosures or access to live conductors which would affect its effectiveness.

IT SHOULD BE NOTED THAT AN RCD MUST NOT BE USED AS THE SOLE MEANS OF PROTECTION AGAINST DIRECT CONTACT WITH LIVE PARTS. 412-06-01

h *Protection against indirect contact*

The method of protection against indirect contact must be determined and recorded. For earthed equipotential bonding and automatic disconnection of supply the adequacy of main equipotential bonding and the connection of all protective conductors with the earth is essential. The loss of the earth connections will convert indirect contact to direct in the event of a fault. 712-01-03 (viii)(c) 413 i *Protective devices*

The presence, accessibility, labelling and condition of devices for electrical protection, isolation and switching should be verified. 712-01-03(x) 131-12-01

It should be established that each circuit is adequately protected with the correct type, size and rating of fuse or circuit-breaker. The suitability of each protective and monitoring device and its overload rating or setting should be checked. 712-01-03(xii)

It must be ascertained that each protective device is correctly located and appropriate to the type of earthing system and to the circuits protected.

Each device for protection, isolation and switching should be readily accessible for normal operation, maintenance and inspection, and be suitably labelled where necessary. 712-01-03(xv) 712-01-03(xiii)

It should be established that a means of emergency switching, or where appropriate emergency stopping, is provided where required by the Regulations, so that the supply can be cut off rapidly to prevent or remove danger. Where a risk of electric shock is involved then the emergency switching device shall switch all the live conductors except as permitted by the Regulations. 712-01-03(x)

When conditions permit, an internal inspection should be made of any emergency switching device and tests should be carried out as described in Section 3.10.3.

j *Enclosures and mechanical protection*

The enclosure and mechanical protection of all electrical equipment should be inspected to ensure that they remain adequate for the type of protection intended. All secondary barriers (to IP2X or IPXXB) should be in place. 713-07 131-05 712-01-03(xiv) 412-03-04

514
608

k *Marking and Labelling*

The labelling of each circuit should be verified. 514-01-01 514-08-01

It should be established that adjacent to every fuse or circuit-breaker there is a label correctly indicating the size and type of the fuse, nominal current of the circuit-breaker and identification of the protected circuit.

It should be ascertained that all switching devices are correctly labelled and identify the circuits controlled.

Notices or labels are required at the following points and equipment within an installation:

(i) at the origin of every installation 514-12-01

a notice of such durable material as to be likely to remain easily legible throughout the life of the installation, marked in indelible characters no smaller than the example in BS 7671

IMPORTANT

This installation should be periodically inspected and tested and a report on its condition obtained, as prescribed in BS 7671 Requirements for Electrical Installations published by the Institution of Electrical Engineers.

Date of last inspection...................................

Recommended date of next inspection...................

shall be fixed in a prominent position at or near the origin of the installation.

(ii) where different voltages are present 514-10

(a) equipment or enclosures within which a nominal voltage exceeding 230 V exists and where the presence of such a voltage would not normally be expected

(b) where terminals or other fixed live parts between which a nominal voltage exceeding 230 V exists are housed in separate enclosures or items of equipment which, although separated, can be reached simultaneously by a person

(c) means of access to all live parts of switchgear and other fixed live parts where different nominal voltages exist.

(iii) earthing and bonding connections

(a) a permanent label to BS 951 with the words 514-13-01

Safety Electrical Connection — Do Not Remove

shall be permanently fixed in a visible position at or near:

1) the point of connection of every earthing conductor to an earth electrode, and
2) the point of connection of every bonding conductor to an extraneous-conductive-part, and
3) the main earth terminal, where separate from main switchgear.

(b) In an ***earth-free*** situation all exposed metalwork must be bonded together but not to the main earthing system. In this situation a suitable permanent notice durably marked in legible type and no smaller than the example in BS 7671 shall be permanently fixed in a visible position. 514-13-02

The equipotential protective bonding conductors associated with the electrical installation in this location **MUST NOT BE CONNECTED TO EARTH.** Equipment having exposed-conductive-parts connected to earth must not be brought into this location.

(iv) Residual Current Devices (RCDs)

Where RCDs are fitted within an installation a suitable permanent notice durably marked in legible type and no smaller than the example in BS 7671 shall be permanently fixed in a prominent position at or near the main distribution board. 514-12-02

This installation, or part of it, is protected by a device which automatically switches off the supply if an earth fault develops. Test quarterly by pressing the button marked 'T' or 'Test'. The device should switch off the supply and should then be switched on to restore the supply. If the device does not switch off the supply when the button is pressed, seek expert advice.

(v) Caravan installations

All caravans and motor caravans shall have a notice fixed near the main switch giving instructions on the connection and disconnection of the caravan installation to the electricity supply. 608-07-05

The notice shall be of a durable material permanently fixed, and bearing in indelible and easily legible characters the text shown in BS 7671.

INSTRUCTIONS FOR ELECTRICITY SUPPLY

TO CONNECT

1. Before connecting the caravan installation to the mains supply, check that:
   (a) the supply available at the caravan pitch supply point is suitable for the caravan electrical installation and appliances, and
   (b) the caravan main switch is in the OFF position.
2. Open the cover to the appliance inlet provided at the caravan supply point and insert the connector of the supply flexible cable.
3. Raise the cover of the electricity outlet provided on the pitch supply point and insert the plug of the supply cable.

THE CARAVAN SUPPLY FLEXIBLE CABLE MUST BE FULLY UNCOILED TO AVOID DAMAGE BY OVERHEATING

4. Switch on at the caravan main switch.
5. Check the operation of residual current devices, if any, fitted in the caravan by depressing the test buttons.

IN CASE OF DOUBT, OR IF AFTER CARRYING OUT THE ABOVE PROCEDURE THE SUPPLY DOES NOT BECOME AVAILABLE, OR IF THE SUPPLY FAILS, CONSULT THE CARAVAN PARK OPERATOR OR THE OPERATOR'S AGENT OR A QUALIFIED ELECTRICIAN

TO DISCONNECT

6. Switch off at the caravan main isolating switch, unplug both ends of the cable.

PERIODIC INSPECTION

Preferably not less than once every three years and more frequently if the vehicle is used more than normal average mileage for such vehicles, the caravan electrical installation and supply cable should be inspected and tested and a report on their condition obtained as prescribed in BS 7671 (formerly the Regulations for Electrical Installations) published by the Institution of Electrical Engineers.

I External influences

Note should be made of any known changes in external influences, building structure, and alterations or additions which may have affected the suitability of the wiring for its present load and method of installation. 712-01-02

Note should be taken of any alterations or additions of an irregular nature to the installation. If unsuitable material has been used the Report should indicate this together with reference to any evident faulty workmanship or design.

3.10 Periodic testing

3.10.1 Periodic testing general

731 Periodic testing is supplementary to the inspection of the installation described in Section 3.8. 731-01-03

The same range and level of testing as for an initial inspection and test is not necessarily required, or indeed possible. Installations that have been previously tested and for which there are comprehensive records of test results may not need the same degree of testing as installations for which there are no records of testing.

Periodic testing may cause danger if the correct procedures are not applied. Persons carrying out periodic testing must be competent in the use of the instruments employed and have adequate knowledge and experience of the type of installation being tested in order to prevent, so far as is reasonably practicable, danger.

Sample testing may be carried out, with the percentage being at the discretion of the tester (see Note 2 of Table 3.3).

Wherever a sample test indicates results significantly different from those previously recorded investigation is necessary. Unless the reason for the difference can be clearly identified as relating only to the sample tested, the size of the sample should be increased. If, in the increased sample, further failures to comply with the requirements of the Regulations are indicated, a 100 per cent test should be made.

3.10.2 Tests to be made

The tests considered appropriate by the person carrying out the inspection should be carried out in accordance with the recommendations in Table 3.3. Reference methods of testing are provided in Section 2.7 of this Guidance Note, but alternative methods which give no less effective results may be used.

TABLE 3.3 - Testing to be carried out where practicable on existing installations (see Notes 1 and 2)

Test (Note 3)	Recommendation
Protective conductors continuity	Between the earth terminal of distribution boards to the following exposed-conductive-parts: • socket-outlet earth connections **(Note 4)** • accessible exposed-conductive-parts of current-using equipment and accessories **(Notes 4** and **5).**
Bonding conductors continuity	• all main bonding conductors • all necessary supplementary bonding conductors.
Ring circuit continuity	Where there are proper records of previous tests, this test may not be necessary. This test should be carried out where inspection/ documentation indicate that there may have been changes made to the ring final circuit.
Insulation resistance	If tests are to be made: • between live conductors, with phase(s) and neutral connected together, and Earth at all final distribution boards • at main and sub-main distribution panels, with final circuit distribution boards isolated from mains. **(Note 6)**
Polarity	At the following positions: • origin of the installation • distribution boards • accessible socket-outlets • extremity of radial circuits. **(Note 7)**
Earth electrode resistance	Test each earth rod or group of rods separately, with the test links removed, and with the installation isolated from the supply source.
Earth fault loop impedance	At the following positions: • origin of the installation • distribution boards • accessible socket-outlets • extremity of radial circuits. **(Note 8)**
Functional tests	
RCDs	Tests as required by Regulation 713-13-01, followed by operation of the functional test button.
Circuit-breakers, isolators and switching devices	Manual operation to prove that the devices disconnect the supply.

Notes:

1) The person carrying out the testing is required to decide which of the above tests are appropriate by using their experience and knowledge of the installation being inspected and tested and by consulting any available records.
2) Where sampling is applied, the percentage used is at the discretion of the tester. However, a percentage of less than 10% is inadvisable.
3) The tests need not be carried out in the order shown in the table.
4) The earth fault loop impedance test may be used to confirm the continuity of protective conductors at socket-outlets and at accessible exposed-conductive-parts of current-using equipment and accessories.
5) Generally, accessibility may be considered to be within 3 metres from the floor or from where a person can stand.
6) Where the circuit includes SPDs or other electronic devices which require a connection to Earth for functional purposes, these devices will require disconnecting to avoid influencing the test result and to avoid damaging them.
7) Where there are proper records of previous tests, this test may not be necessary.
8) Some earth loop impedance testers may trip RCDs in the circuit.

3.10.3 Detailed periodic testing

713 a *Continuity of protective conductors and equipotential bonding conductors*

If an electrical installation is isolated from the supply it is permissible to disconnect protective and equipotential bonding conductors from the main earthing terminal in order to verify their continuity. 713-02

Where an electrical installation cannot be isolated from the supply the protective and equipotential bonding conductors must **NOT** be disconnected as, under fault conditions, the exposed and extraneous-conductive-parts could be raised to a dangerous level above earth potential.

The use of an earth fault loop impedance tester is often the most convenient way of continuity testing.

When testing the effectiveness of main equipotential bonding conductors, the resistance value between a service pipe or other extraneous-conductive-part and the main earthing terminal should be of the order of 0.05 Ω or less.

Supplementary bonding conductors should similarly have a resistance of 0.05 Ω or less.

b *Insulation resistance*

Insulation resistance tests should be made on electrically isolated circuits with any electronic equipment which might be damaged by application of the test voltage disconnected or only a measurement to protective earth made, with the phase and neutral connected together. 713-04

Check that information/warnings are given at the distribution board of circuits or equipment likely to be damaged by testing. Any diagram, chart or table should also include this warning.

Where practicable, the tests may be applied to the whole of the installation with all fuse links in place and all switches closed. Alternatively, the installation may be tested in parts, one distribution board at a time. See Table 3.3.

A d.c. voltage not less than that stated in Table 2.2 should be employed.

Table 2.2 of this Guide and Table 71A of BS 7671 require a minimum insulation resistance of 0.5 MΩ; however, if the resistance is less than 2 MΩ, further investigation is required to determine the cause of the low reading.

Where equipment is disconnected for these tests and the equipment has exposed-conductive-parts required by the Regulations to be connected to protective conductors, the insulation resistance between the exposed-conductive-parts and all live parts of the equipment should be measured separately and should comply with the requirements of the appropriate British Standard for the equipment. If there is no appropriate British Standard for the equipment the insulation resistance should be not less than 0.5 megohm.

c *Polarity*

Tests should be made to verify that:

(i) the polarity is correct at the meter and distribution board 713-09

(ii) every fuse and single-pole control and protective devices are connected in phase conductors only

(iii) conductors are correctly connected to socket-outlets and other accessories/equipment

(iv) except for E14 and E27 lampholders to BS EN 60238, centre-contact bayonet and Edison screw lampholders have their outer or screw contacts connected to the earthed neutral conductor

(v) all multi-pole devices are correctly installed.

It should be established whether there have been any alterations or additions to the installation since its last inspection and test. If there have been no alterations or additions then sample tests should be made of at least 10 per cent of all single-pole and multi-pole control devices and of any centre-contact lampholders, together with 100 per cent of socket-outlets. If any incorrect polarity is found then a full test should be made in that part of the installation supplied by the particular distribution board concerned, and the sample testing increased to 25 per cent for the remainder of the installation. If additional cases of incorrect polarity are found in the 25 per cent sample a full test of the complete installation should be made.

d *Earth fault loop impedance*

Where protective measures are used which require a knowledge of earth fault loop impedance, the relevant impedance should be measured, or determined by an equally effective method.

Earth fault loop impedance tests should be carried out at the locations indicated below:

(i) origin of the installation

(ii) distribution boards

(iii) accessible socket-outlets

(iv) extremity of radial circuits.

Motor circuits

Loop impedance tests on motor circuits can only be carried out on the supply side of isolated motor controlgear. Continuity tests between the circuit protective conductor and motor are then necessary.

Circuits incorporating an RCD

Where the installation incorporates an RCD, the value of earth fault loop impedance obtained in the test should be related to the rated residual operating current ($I_{\Delta n}$) of the protective device, to verify compliance with Section 413, of BS 7671.

$Z_S \, I_{\Delta n} \leq 50$ V for TN systems 413-02-16

$R_A \, I_{\Delta n} \leq 50$ V for TT systems. 413-02-20

Where items of stationary equipment having a protective conductor current exceeding 3.5 mA in normal service are supplied from an installation forming part of a TT system, the product of the total protective conductor current (in amperes) and twice the resistance of the installation earth electrode(s) (in ohms) must not exceed 50. 607-05-01

e *Operation of residual current devices*

Where there is RCD protection, the effective operation of each RCD must be verified by a test simulating an appropriate fault condition independent of any test facility incorporated in the device, followed by operation of the integral test device. The nominal rated tripping current for protection of a socket-outlet for use with equipment outdoors or installed outside the general installation equipotential zone must not exceed 30 mA. 713-13

Refer to Section 2.7.16 for detailed test procedure for RCDs.

f *Operation of overcurrent circuit-breakers*

Where protection against overcurrent is provided by circuit-breakers, the manual operating mechanism of each circuit-breaker should be operated to verify that the device opens and closes satisfactorily. 713-13

It is not normally necessary or practicable to test the operation of the automatic tripping mechanism of circuit-breakers. Any such test would need to be made at a current substantially exceeding the minimum tripping current in order to achieve operation within a reasonable time. For circuit-breakers to BS EN 60898 a test current of not less than two and a half times the nominal rated tripping current of the device is needed for operation within 1 minute, and much larger test currents are necessary to verify operation of the mechanism for instantaneous tripping. 712-01-03 (xii)

For circuit-breakers of the sealed type, designed not to be maintained, if there is doubt about the integrity of the automatic mechanism it will normally be more convenient to replace the device than to make further tests. Such doubt may arise from visual inspection, if the device appears to have suffered damage or undue deterioration, or where there is evidence that the device may have failed to operate satisfactorily in service.

Circuit-breakers with the facility for injection testing should be so tested and, if appropriate, relay settings confirmed. 712-01-03(x)

g Operation of devices for isolation and switching

Where means are provided in accordance with the requirements of the Regulations for isolation and switching, they should be operated to verify their effectiveness and checked to ensure adequate and correct labelling. 514-01-01

Easy access to such devices must be maintained, and effective operation must not be impaired by any material placed near the device. Access and operation areas may be required to be marked to ensure they are kept clear.

For isolating devices in which the position of the contacts or other means of isolation is externally visible, visual inspection of operation is sufficient and no test need be made.

The operation of every safety switching device should be checked by operating the device in the manner normally intended to confirm that it performs its function correctly in accordance with the requirements of BS 7671.

Where it is a requirement that the device interrupts all the supply conductors, the use of a test lamp or instrument connected between each phase and a connection to neutral on the load side of the switching device may be necessary. Reliance should not be placed on a simple observation that the equipment controlled has ceased to operate.

Where switching devices are provided with detachable or lockable handles in accordance with the Regulations, a check should be made to verify that the handles or keys are not interchangeable with others available within the premises.

Where any form of interlocking is provided, e.g. between a main circuit-breaker and an outgoing switch or isolation device, the integrity of the interlocking must be verified. 713-13-01

Where switching devices are provided for isolation or for mechanical maintenance switching, the integrity of the means provided to prevent any equipment from being unintentionally or inadvertently energised or reactivated must be verified. 461-01-02 462-01-03

Note
An inspection checklist which may be used when carrying out periodic inspections is included in Section 2.6.3 of this Guidance Note.

3.10.4 Periodic Inspection Report

741 BS 7671 requires that the results and extent of periodic inspection and 741-01-02
744 testing shall be recorded on a Periodic Inspection Report and provided 744-01-01
to the person ordering the inspection.

The report must include: 741-01-02

(i) a description of the extent of the work, including the parts of the installation inspected and details of what the inspection and testing covered

(ii) any limitations which may have been imposed during the inspection and testing of the installation

(iii) details of any damage, deterioration, defects and dangerous conditions and any non-compliance with BS 7671 which may give rise to danger 744-01-02

(iv) schedule of inspections

(v) schedule of test results.

Any immediately dangerous condition should preferably be rectified. If not, the defect should be reported in writing without delay to the employer or responsible employee (see Regulation 3 of the Electricity at Work Regulations).

Installations including those constructed in accordance with earlier 731-01-04
editions of BS 7671 should be inspected and tested for compliance with the current edition of BS 7671 and departures recorded. However, reference should be made to the note by the Health and Safety Executive following the preface to BS 7671, that installations conforming to earlier editions and not complying with the current edition do not necessarily fail to achieve conformity with the Electricity at Work Regulations 1989.

Guidance on the action to be taken is to be given in the Observations and Recommendations section of the Periodic Inspection Report by numbering each observation (non-compliance) 1 to 4.

If the number 1 is allocated to the installation, indicating that it requires urgent attention, then the overall assessment must be that the installation is unsatisfactory. An example of this is an installation which has no earth. If numbers 2, 3 or 4 are allocated, the person carrying out the test will have to use judgement to determine whether or not the installation can be classed as satisfactory.

Section 4 — Test Instruments

4.1 Instrument standard

BS EN 61010 : Safety requirements for electrical equipment for measurement control and laboratory use is the basic safety standard for electrical test instruments.

The basic instrument standard is BS EN 61557: Electrical safety in low voltage distribution systems up to 1000 V a.c. and 1500 V d.c. Equipment for testing, measuring or monitoring of protective measures. This standard includes performance requirements and requires compliance with BS EN 61010.

In Section 1.1 Safety, reference was made to the use of test leads which conform to HSE Guidance Note GS38. The safety measures and procedures set out in HSE Guidance Note GS38 should be observed for all instruments, leads, probes and accessories. It should be noted that some test instrument manufacturers' advise that their instruments be used in conjunction with fused test leads and probes. Other manufacturers advise the use of non-fused leads and probes when the instrument has in-built electrical protection but it should be noted that such electrical protection does not extend to the probes and leads.

4.2 Instrument accuracy

A basic measurement accuracy of 5 % is usually adequate for these test instruments. In the case of analogue instruments, a basic accuracy of 2 % of full scale deflection will provide the required accuracy measurement over a useful proportion of the scale.

It should not be assumed that the accuracy of the reading taken in normal field use will be as good as the basic accuracy. The 'operating accuracy' is always worse than the basic accuracy, and additional errors derive from three sources:

(i) *Instrument errors:* basic instrument accuracy applies only in ideal conditions; the actual reading accuracy will also be affected by the operator's ability, battery condition, generator cranking speed, ambient temperature and orientation of the instrument

(ii) *Loss of calibration:* instruments should be regularly recalibrated using standards traceable to National Standards, and should be checked after any mechanical or electrical mishandling

(iii) *Field errors:* the instrument reading accuracy will also be affected by external influences as a result of working in the field environment. These influences may be in many forms, and some sources of such inaccuracies are described in the appropriate sections.

BS EN 61557 requires a maximum operating error of 30% of reading over the stated measurement range.

To achieve satisfactory in-service performance, it is essential to be fully informed about the test equipment, how it is to be used, and the accuracy to be expected.

4.3 Low-resistance ohmmeters

The instrument used for low-resistance tests may be either a specialised low-resistance ohmmeter, or the continuity range of an insulation and continuity tester. The test current may be d.c. or a.c. It is recommended that it be derived from a source with no-load voltage between 4 V and 24 V, and a short-circuit current not less than 200 mA. 713-02-01

The measuring range should cover the span 0.2 ohm to 2 ohms, with a resolution of at least 0.01 ohm for digital instruments.

Field effects contributing to in-service errors are contact resistance, test lead resistance, a.c. interference and thermocouple effects in mixed metal systems.

Whilst contact resistance cannot be eliminated with two-terminal testers, and can introduce errors, the effects of lead resistance can be eliminated by measuring this prior to a test, and subtracting the resistance from the final value. Interference from an external a.c. source (interference pickup) cannot be eliminated, although it may be indicated by vibration of the pointer of an analogue instrument. Thermocouple effects can be eliminated by reversing the test probes and averaging the resistance readings taken in each direction.

Instruments to BS EN 61557-4 will meet the above requirements.

4.4 Insulation resistance ohmmeters

The instrument used should be capable of developing the test voltage required across the load.

The test voltage required is:

(i) 250 V d.c. for SELV and PELV circuits Table 71A

(ii) 500 V d.c. for all circuits rated up to and including 500 V, but excluding extra-low voltage circuits mentioned above

(iii) 1000 V d.c. for circuits rated above 500 V up to 1000 V.

The tester must be capable of supplying an output current of 1 mA at the required test voltage. 713-04-03

The factors affecting in-service reading accuracy include 50 Hz currents induced into cables under test, and capacitance in the test object. These errors cannot be eliminated by test procedures. Capacitance may be as high as 5 μF, and the instrument should have an automatic discharge facility capable of safely discharging such a capacitance. Following an insulation resistance test, the instrument should be left connected until the capacitance within the installation has fully discharged.

Instruments conforming to BS EN 61557-2 will fulfil all the above instrument requirements.

4.5 Applied voltage testers

For the purposes of the applied voltage tests, the instrument used should be capable of steadily increasing the test voltage with the accuracy of the indication of the output voltage being ±5%. It should have means of indicating a failure of insulation, and should be capable of maintaining the test voltage continuously for at least 1 minute.

Care should be taken to prevent bare exposed metalwork becoming live unnecessarily, and the integrity of the earth connection of the test equipment should be verified.

Unless specifically required to overcome the capacitance in larger equipment or leakage that might be permissible with non-conducting floors and walls, the maximum output current of the tester should not exceed 5 mA. In general, the maximum test voltage required by BS 7671 is 3750 V a.c. rms. If the maximum output of the test equipment can 711-01-01
exceed 5 mA, it must be regarded as lethal and precautions must be 731-01-05
taken to prevent danger to the person carrying out the test, and to anyone who might be near the apparatus or installation being tested.

4.6 Earth fault loop impedance testers

These instruments operate by circulating a current from the phase conductor into the protective earth. This will raise the potential of the protective earth system.

To minimise electric shock hazard from the potential of the protective conductor, the test duration should be within safe limits. This means that the instrument should cut off the test current after 40 ms or a time determined by the safety limits of IEC 479, if the voltage rise of the protective conductor exceeds 50 V during the test.

Instrument accuracy decreases as scale reading reduces. Aspects affecting in-service reading accuracy include transient variations of mains voltage during the test period, mains interference, test lead resistance and errors in impedance measurement as a result of the test method. To allow for the effect of transient voltages the test should be repeated at least once. The other effects cannot be eliminated by test procedures.

For circuits rated up to 50 A, a phase-earth loop tester with a resolution of 0.01 ohm should be adequate. In general, such instruments can be relied upon to be accurate down to values of around 0.2 ohm.

Instruments conforming to BS EN 61557-3 will fulfil the above requirements.

These instruments may also offer additional facilities for measuring prospective short-circuit current. The basic measuring principle is generally the same as for earth fault loop impedance testers. The current is calculated by dividing the earth fault loop impedance value into the mains voltage. Instrument accuracy is determined by the same factors as for loop testers. In this case instrument accuracy decreases as scale reading increases, because the loop value is divided into the mains voltage. It is important to note these aspects and the manufacturer's documentation should be referred to.

4.7 Earth electrode resistance testers

This may be a 4-terminal instrument (or a-3 terminal one where a combined lead to the earth electrode would not have a significant resistance compared with the electrode resistance) so that the resistance of the test leads and temporary spike resistance can be eliminated from the test result.

Aspects affecting in-service reading accuracy include the effects of temporary spike resistance, interference currents and the layout of the test electrodes. The instrument should carry some facility to check that the resistance to earth of the temporary potential and current spikes are within the operating limits of the instrument. It may be helpful to note that instruments complying with BS EN 61557-5 incorporate this facility. Care should be exercised to ensure that temporary spikes are positioned with reasonable accuracy.

4.8 RCD testers

The test instrument should be capable of applying the full range of test current to an in-service accuracy as given in BS EN 61557-6. This in-service reading accuracy will include the effects of voltage variations around the nominal voltage of the tester.

To check RCD operation and to minimise danger during the test, the test current should be applied for no longer than 2 s.

Instruments conforming to BS EN 61557-6 will fulfil the above requirements.

Section 5 — Forms

5.1 Initial inspection and testing

Forms 1 to 4 are designed for use when inspecting and testing a new installation, or an alteration or addition to an existing installation. The forms comprise the following:

1. Short form of Electrical Installation Certificate (to be used when one person is responsible for the design, construction, inspection and testing of an installation)
2. Electrical Installation Certificate (Standard form from Appendix 6 of BS 7671)
3. Schedule of Inspections
4. Schedule of Test Results.

Notes on completion and guidance for recipients is provided with the form.

5.2 Minor works

The complete set of forms for initial inspection and testing may not be appropriate for minor works. When an addition to an electrical installation does not extend to the installation of a new circuit, the minor works form may be used. This form is intended for such work as the addition of a socket-outlet or lighting point to an existing circuit, or for repair or modification.

Form 5 is the Minor Electrical Installation Works Certificate from Appendix 6 of BS 7671.

Notes on completion and guidance for recipients is provided with the form.

5.3 Periodic inspection

Form 6, the Periodic Inspection Report from Appendix 6 of BS 7671, is for use when carrying out routine periodic inspection and testing of an existing installation. It is not for use when alterations or additions are made. A Schedule of Inspections (3) and Schedule of Test Results (4) should accompany the Periodic Inspection Report (6).

Notes on completion and guidance for recipients is provided with the form.

CERTIFICATION AND REPORTING

The introduction to Appendix 6 of BS 7671 : 2001 (Model forms for certification and reporting) is reproduced on this page.

Introduction

(i) The Electrical Installation Certificate required by Part 7 of BS 7671 shall be made out and signed or otherwise authenticated by a competent person or persons in respect of the design, construction, inspection and testing of the work.

(ii) The Minor Works Certificate required by Part 7 of BS 7671 shall be made out and signed or otherwise authenticated by a competent person in respect of the inspection and testing of an installation.

(iii) The Periodic Inspection Report required by Part 7 of BS 7671 shall be made out and signed or otherwise authenticated by a competent person in respect of the inspection and testing of an installation.

(iv) Competent persons will, as appropriate to their function under (i) (ii) and (iii) above, have a sound knowledge and experience relevant to the nature of the work undertaken and to the technical standards set down in this British Standard, be fully versed in the inspection and testing procedures contained in this Standard and employ adequate testing equipment.

(v) Electrical Installation Certificates will indicate the responsibility for design, construction, inspection and testing, whether in relation to new work or further work on an existing installation.

Where design, construction and inspection and testing is the responsibility of one person a Certificate with a single signature declaration in the form shown below may replace the multiple signatures section of the model form.

FOR DESIGN, CONSTRUCTION, INSPECTION & TESTING.

I being the person responsible for the Design, Construction, Inspection & Testing of the electrical installation (as indicated by my signature below), particulars of which are described above, having exercised reasonable skill and care when carrying out the Design, Construction, Inspection & Testing, hereby CERTIFY that the said work for which I have been responsible is to the best of my knowledge and belief in accordance with BS 7671 : ...., amended to ....(date) except for the departures, if any, detailed as follows.

(vi) A Minor Works Certificate will indicate the responsibility for design, construction, inspection and testing of the work described in Part 4 of the certificate.

(vii) A Periodic Inspection Report will indicate the responsibility for the inspection and testing of an installation within the extent and limitations specified on the report.

(viii) A schedule of inspections and a schedule of test results as required by Part 7 (of BS 7671) shall be issued with the associated Electrical Installation Certificate or Periodic Inspection Report.

(ix) When making out and signing a form on behalf of a company or other business entity, individuals shall state for whom they are acting.

(x) Additional forms may be required as clarification, if needed by non-technical persons, or in expansion, for larger or more complex installations.

(xi) The IEE Guidance Note 3 provides further information on inspection and testing on completion and for periodic inspections.

ELECTRICAL INSTALLATION CERTIFICATES
NOTES FOR FORMS 1 AND 2

1. The Electrical Installation Certificate is to be used only for the initial certification of a new installation or for an alteration or addition to an existing installation where new circuits have been introduced.

   It is not to be used for a Periodic Inspection for which a Periodic Inspection Report form should be used. For an alteration or addition which does not extend to the introduction of new circuits, a Minor Electrical Installation Works Certificate may be used.

   The original Certificate is to be given to the person ordering the work (Regulation 742-01-03). A duplicate should be retained by the contractor.

2. This Certificate is only valid if accompanied by the Schedule of Inspections and the Schedule(s) of Test Results.

3. The signatures appended are those of the persons authorised by the companies executing the work of design, construction and inspection and testing respectively. A signatory authorised to certify more than one category of work should sign in each of the appropriate places.

4. The time interval recommended before the first periodic inspection must be inserted (see IEE Guidance Note 3 for guidance).

5. The page numbers for each of the Schedules of Test Results should be indicated, together with the total number of sheets involved.

6. The maximum prospective fault current recorded should be the greater of either the short-circuit current or the earth fault current.

7. The proposed date for the next inspection should take into consideration the frequency and quality of maintenance that the installation can reasonably be expected to receive during its intended life, and the period should be agreed between the designer, installer and other relevant parties.

ELECTRICAL INSTALLATION CERTIFICATE (notes 1 and 2)

(REQUIREMENTS FOR ELECTRICAL INSTALLATIONS - BS 7671 [IEE WIRING REGULATIONS])

DETAILS OF THE CLIENT (note 1)

....

....

....

INSTALLATION ADDRESS

....

....

.... Postcode ....

DESCRIPTION AND EXTENT OF THE INSTALLATION Tick boxes as appropriate

Description of installation: ....

Extent of installation covered by this Certificate: ....

....

....

....

....

....

....

New installation ☐

Addition to an existing installation ☐

Alteration to an existing installation ☐

FOR DESIGN, CONSTRUCTION, INSPECTION & TESTING

I being the person responsible for the Design, Construction, Inspection & Testing of the electrical installation (as indicated by my signature below), particulars of which are described above, having exercised reasonable skill and care when carrying out the Design, Construction, Inspection & Testing, hereby CERTIFY that the said work for which I have been responsible is to the best of my knowledge and belief in accordance with BS 7671 : ...., amended to .... (date) except for the departures, if any, detailed as follows:

Details of departures from BS 7671 (Regulations 120-01-03, 120-02):

The extent of liability of the signatory is limited to the work described above as the subject of this Certificate.

Name (IN BLOCK LETTERS):.. Position: ....

Signature (note 3):.. Date: ....

For and on behalf of: ....

Address: ....

....

.... Postcode....................... Tel No: ....

NEXT INSPECTION

I recommend that this installation is further inspected and tested after an interval of not more than .... years/months (notes 4 and 7)

SUPPLY CHARACTERISTICS AND EARTHING ARRANGEMENTS Tick boxes and enter details, as appropriate

Earthing arrangements	**Number and Type of Live Conductors**		**Nature of Supply Parameters**	**Supply Protective Device Characteristics**
TN-C ☐	a.c. ☐	d.c. ☐	Nominal voltage, U/U_o (1) V	Type: ....
TN-S ☐	1-phase, 2-wire ☐	2-pole ☐	Nominal frequency, f (1) ....Hz	....
TN-C-S ☐	1-phase, 3-wire ☐	3-pole ☐	Prospective fault current, I_{pf} (2) ....kA (note 6)	Nominal current rating ....A
TT ☐	2-phase, 3-wire ☐	other ☐	External loop impedance, Z_e (2) ....Ω	
IT ☐	3-phase, 3-wire ☐		*(Note: (1) by enquiry, (2) by enquiry or by measurement)*	
Alternative source ☐ of supply (to be detailed on attached schedules)	3-phase, 4-wire ☐			

PARTICULARS OF INSTALLATION REFERRED TO IN THE CERTIFICATE
Tick boxes and enter details, as appropriate

Means of Earthing	**Maximum Demand**		
Supplier's facility ☐	Maximum demand (load) .... Amps per phase		
	Details of Installation Earth Electrode (*where applicable)*		
Installation earth electrode ☐	Type (e.g. rod(s), tape etc)	Location	Electrode resistance to earth
	....	....	.... Ω

Main Protective Conductors

Earthing conductor: material .... csa ....mm² connection verified ☐

Main equipotential bonding conductors material .... csa ....mm² connection verified ☐

To incoming water and/or gas service ☐ To other elements ....

Main Switch or Circuit-breaker

BS, Type.................................... No. of poles .... Current rating ....A Voltage rating V

Location .... Fuse rating or setting ....A

Rated residual operating current $I_{\Delta n}$ = ... mA, and operating time of .... ms (at $I_{\Delta n}$) (applicable only where an RCD is suitable and is used as a main circuit-breaker)

COMMENTS ON EXISTING INSTALLATION: (In the case of an alteration or additions see Section 743)

....

....

....

....

....

....

....

....

SCHEDULES (note 2)

The attached Schedules are part of this document and this Certificate is valid only when they are attached to it.

.... Inspection Schedules and .... Test Result Schedules are attached.

(Enter quantities of schedules attached).

GUIDANCE FOR RECIPIENTS

This safety Certificate has been issued to confirm that the electrical installation work to which it relates has been designed, constructed and inspected and tested in accordance with British Standard 7671 (The IEE Wiring Regulations).

You should have received an original Certificate and the contractor should have retained a duplicate Certificate. If you were the person ordering the work, but not the user of the installation, you should pass this Certificate, or a full copy of it including the schedules, immediately to the user.

The "original" Certificate should be retained in a safe place and be shown to any person inspecting or undertaking further work on the electrical installation in the future. If you later vacate the property, this Certificate will demonstrate to the new owner that the electrical installation complied with the requirements of British Standard 7671 at the time the Certificate was issued. The Construction (Design and Management) Regulations require that for a project covered by those Regulations, a copy of this Certificate, together with schedules is included in the project health and safety documentation.

For safety reasons, the electrical installation will need to be inspected at appropriate intervals by a competent person. The maximum time interval recommended before the next inspection is stated on Page 1 under "Next Inspection".

This Certificate is intended to be issued only for a new electrical installation or for new work associated with an alteration or addition to an existing installation. It should not have been issued for the inspection of an existing electrical installation. A "Periodic Inspection Report" should be issued for such a periodic inspection.

Form No /2

ELECTRICAL INSTALLATION CERTIFICATE (notes 1 and 2)

(REQUIREMENTS FOR ELECTRICAL INSTALLATIONS - BS 7671 [IEE WIRING REGULATIONS])

DETAILS OF THE CLIENT (note 1)

....

....

....

INSTALLATION ADDRESS

....

....

.... Postcode ....

DESCRIPTION AND EXTENT OF THE INSTALLATION Tick boxes as appropriate (note 1)

Description of installation: ....

Extent of installation covered by this Certificate:

....

....

....

....

....

....

New installation ☐

Addition to an existing installation ☐

Alteration to an existing installation ☐

FOR DESIGN

I/We being the person(s) responsible for the design of the electrical installation (as indicated by my/our signatures below), particulars of which are described above, having exercised reasonable skill and care when carrying out the design hereby CERTIFY that the design work for which I/we have been responsible is to the best of my/our knowledge and belief in accordance with BS 7671 : ...., amended to...............................(date) except for the departures, if any, detailed as follows:

Details of departures from BS 7671 (Regulations 120-01-03, 120-02):

The extent of liability of the signatory or the signatories is limited to the work described above as the subject of this Certificate.

For the DESIGN of the installation: **(Where there is mutual responsibility for the design)

Signature: .... Date:............................ Name (BLOCK LETTERS):... Designer No 1

Signature:... Date:............................ Name (BLOCK LETTERS):... Designer No 2**

FOR CONSTRUCTION

I/We being the person(s) responsible for the construction of the electrical installation (as indicated by my/our signatures below), particulars of which are described above, having exercised reasonable skill and care when carrying out the construction hereby CERTIFY that the construction work for which I/we have been responsible is to the best of my/our knowledge and belief in accordance with BS 7671 : ...., amended to ....(date) except for the departures, if any, detailed as follows:

Details of departures from BS 7671 (Regulations 120-01-03, 120-02):

The extent of liability of the signatory is limited to the work described above as the subject of this Certificate.

For CONSTRUCTION of the installation:

Signature .... Date...

Name (BLOCK LETTERS) .... Constructor

FOR INSPECTION & TESTING

I/We being the person(s) responsible for the inspection & testing of the electrical installation (as indicated by my/our signatures below), particulars of which are described above, having exercised reasonable skill and care when carrying out the inspection & testing hereby CERTIFY that the work for which I/we have been responsible is to the best of my/our knowledge and belief in accordance with BS 7671 : ...., amended to ....(date) except for the departures, if any, detailed as follows:

Details of departures from BS 7671 (Regulations 120-01-03, 120-02):

The extent of liability of the signatory is limited to the work described above as the subject of this Certificate.

For INSPECTION AND TEST of the installation:

Signature .... Date...

Name (BLOCK LETTERS) .... Inspector

NEXT INSPECTION (notes 4 and 7)

I/We the designer(s), recommend that this installation is further inspected and tested after an interval of not more than .... years/months.

Page 1 of (note 5)

PARTICULARS OF SIGNATORIES TO THE ELECTRICAL INSTALLATION CERTIFICATE (note 3)

Designer (No 1)
Name: .... Company: ....
Address: ....
.... Postcode: .... Tel No: ....

Designer (No 2)
(if applicable)
Name: .... Company: ....
Address: ....
.... Postcode: .... Tel No: ....

Constructor
Name: .... Company: ....
Address: ....
.... Postcode: .... Tel No: ....

Inspector
Name: .... Company: ....
Address: ....
.... Postcode: .... Tel No: ....

SUPPLY CHARACTERISTICS AND EARTHING ARRANGEMENTS
Tick boxes and enter details, as appropriate

Earthing arrangements	Number and Type of Live Conductors	Nature of Supply Parameters	Supply Protective Device Characteristics
TN-C ☐	a.c. ☐ d.c. ☐	Nominal voltage, U/U_0 (1) V	Type: ....
TN-S ☐	1-phase, 2-wire ☐ 2-pole ☐	Nominal frequency, f (1) ....Hz	....
TN-C-S ☐	1-phase, 3-wire ☐ 3-pole ☐	Prospective fault current, I_{pf} (2) ....kA (note 6)	Nominal current rating ....A
TT ☐	2-phase, 3-wire ☐ other ☐	External loop impedance, Z_e (2) ....Ω	
IT ☐	3-phase, 3-wire ☐	*(Note: (1) by enquiry, (2) by enquiry or by measurement)*	
Alternative source ☐ of supply (to be detailed on attached schedules)	3-phase, 4-wire ☐		

PARTICULARS OF INSTALLATION REFERRED TO IN THE CERTIFICATE
Tick boxes and enter details, as appropriate

Means of Earthing

Supplier's facility ☐

Installation earth electrode ☐

Maximum Demand

Maximum demand (load) .... Amps per phase

Details of Installation Earth Electrode *(where applicable)*

Type (e.g. rod(s), tape etc)	Location	Electrode resistance to earth
....	....	.... Ω

Main Protective Conductors

Earthing conductor: material .... csa ....mm^2 connection verified ☐

Main equipotential bonding conductors material .... csa ....mm^2 connection verified ☐

To incoming water and/or gas service ☐ To other elements ....

Main Switch or Circuit-breaker

BS, Type.................................... No. of poles.................. Current rating ....A Voltage rating V

Location.. Fuse rating or setting ....A

Rated residual operating current $I_{\Delta n}$ = ... mA, and operating time of .... ms (at $I_{\Delta n}$) (applicable only where an RCD is suitable and is used as a main circuit-breaker)

COMMENTS ON EXISTING INSTALLATION:
(In the case of an alteration or additions see Section 743)

....

....

....

....

SCHEDULES (note 2)

The attached Schedules are part of this document and this Certificate is valid only when they are attached to it.
.... Inspection Schedules and .... Test Result Schedules are attached.
(Enter quantities of schedules attached).

Page 2 of (note 5)

ELECTRICAL INSTALLATION CERTIFICATE
GUIDANCE FOR RECIPIENTS (to be appended to the Certificate)

This safety Certificate has been issued to confirm that the electrical installation work to which it relates has been designed, constructed and inspected and tested in accordance with British Standard 7671 (The IEE Wiring Regulations).

You should have received an original Certificate and the contractor should have retained a duplicate Certificate. If you were the person ordering the work, but not the user of the installation, you should pass this Certificate, or a full copy of it including the schedules, immediately to the user.

The "original" Certificate should be retained in a safe place and be shown to any person inspecting or undertaking further work on the electrical installation in the future. If you later vacate the property, this Certificate will demonstrate to the new owner that the electrical installation complied with the requirements of British Standard 7671 at the time the Certificate was issued. The Construction (Design and Management) Regulations require that for a project covered by those Regulations, a copy of this Certificate, together with schedules is included in the project health and safety documentation.

For safety reasons, the electrical installation will need to be inspected at appropriate intervals by a competent person. The maximum time interval recommended before the next inspection is stated on Page 1 under "Next Inspection".

This Certificate is intended to be issued only for a new electrical installation or for new work associated with an alteration or addition to an existing installation. It should not have been issued for the inspection of an existing electrical installation. A "Periodic Inspection Report" should be issued for such a periodic inspection.

The Certificate is only valid if a Schedule of Inspections and Schedule of Test Result are appended.

Form 3 Form No /3

SCHEDULE OF INSPECTIONS

Methods of protection against electric shock

(a) Protection against both direct and indirect contact:

- [ ] (i) SELV (note 1)
- [ ] (ii) Limitation of discharge of energy

(b) Protection against direct contact: (note 2)

- [ ] (i) Insulation of live parts
- [ ] (ii) Barriers or enclosures
- [ ] (iii) Obstacles (note 3)
- [ ] (iv) Placing out of reach (note 4)
- [ ] (v) PELV
- [ ] (vi) Presence of RCD for supplementary protection

(c) Protection against indirect contact:

(i) EEBADS including:

- [ ] Presence of earthing conductor
- [ ] Presence of circuit protective conductors
- [ ] Presence of main equipotential bonding conductors
- [ ] Presence of supplementary equipotential bonding conductors
- [ ] Presence of earthing arrangements for combined protective and functional purposes
- [ ] Presence of adequate arrangements for alternative source(s), where applicable
- [ ] Presence of residual current device(s)
- [ ] (ii) Use of Class II equipment or equivalent insulation (note 5)
- [ ] (iii) Non-conducting location: (note 6) Absence of protective conductors
- [ ] (iv) Earth-free equipotential bonding: (note 8) Presence of earth-free equipotential bonding conductors
- [ ] (v) Electrical separation (note 8)

Prevention of mutual detrimental influence

- [ ] (a) Proximity of non-electrical services and other influences
- [ ] (b) Segregation of band I and band II circuits or band II insulation used
- [ ] (c) Segregation of safety circuits

Identification

- [ ] (a) Presence of diagrams, instructions, circuit charts and similar information
- [ ] (b) Presence of danger notices and other warning notices
- [ ] (c) Labelling of protective devices, switches and terminals
- [ ] (d) Identification of conductors

Cables and conductors

- [ ] (a) Routing of cables in prescribed zones or within mechanical protection
- [ ] (b) Connection of conductors
- [ ] (c) Erection methods
- [ ] (d) Selection of conductors for current-carrying capacity and voltage drop
- [ ] (e) Presence of fire barriers, suitable seals and protection against thermal effects

General

- [ ] (a) Presence and correct location of appropriate devices for isolation and switching
- [ ] (b) Adequacy of access to switchgear and other equipment
- [ ] (c) Particular protective measures for special installations and locations
- [ ] (d) Connection of single-pole devices for protection or switching in phase conductors only
- [ ] (e) Correct connection of accessories and equipment
- [ ] (f) Presence of undervoltage protective devices
- [ ] (g) Choice and setting of protective and monitoring devices for protection against indirect contact and/or overcurrent
- [ ] (h) Selection of equipment and protective measures appropriate to external influences
- [ ] (i) Selection of appropriate functional switching devices

Inspected by .... Date ....

Notes:

✓ to indicate an inspection has been carried out and the result is satisfactory

✗ to indicate an inspection has been carried out and the result was unsatisfactory

N/A to indicate the inspection is not applicable

1. SELV An extra-low voltage system which is electrically separated from Earth and from other systems. The particular requirements of the Regulations must be checked (see Regulations 411-02 and 471-02)
2. Method of protection against direct contact - will include measurement of distances where appropriate
3. Obstacles - only adopted in special circumstances (see Regulations 412-04 and 471-06)
4. Placing out of reach - only adopted in special circumstances (see Regulations 412-05 and 471-07)
5. Use of Class II equipment - infrequently adopted and only when the installation is to be supervised (see Regulations 413-03 and 471-09)
6. Non-conducting locations - not applicable in domestic premises and requiring special precautions (see Regulations 413-04 and 471-10)
7. Earth-free local equipotential bonding - not applicable in domestic premises, only used in special circumstances (see Regulations 413-05 and 471-11)
8. Electrical separation (see Regulations 413-06 and 471-12)

Page of

Form 4

Form No /4

SCHEDULE OF TEST RESULTS

Contractor: ....

Test Date: ....

Signature ....

Address/Location of distribution board:

....

....

....

* Type of Supply: TN-S/TN-C-S/TT

* Z_e at origin: ....ohms

* PFC: ....kA

Instruments

loop impedance:..

continuity: ....

insulation: ....

RCD tester: ....

Method of protection against indirect contact: ....

Equipment vulnerable to testing: ....

Description of Work: ....

Circuit Description	Overcurrent Device * Short-circuit capacity: ....kA		Wiring Conductors		Test Results										
					Continuity			Insulation Resistance		Polarity	Earth Loop Imped-ance Z_s	Functional Testing		Remarks
	type	Rating I_n A	live mm^2	cpc mm^2	$R_1 + R_2$ Ω	R_2 Ω	Ring	Live/ Live MΩ	Live/ Earth MΩ		Ω	RCD time ms	Other	
1	2	3	4	5	* 6	* 7	* 8	* 9	* 10	* 11	* 12	* 13	* 14	15

Deviations from Wiring Regulations and special notes:

* See notes on schedule of test results

Page of

NOTES ON SCHEDULE OF TEST RESULTS

* **Type of supply** is ascertained from the supply company or by inspection.

* **Z_e at origin.** When the maximum value declared by the electricity supplier is used, the effectiveness of the earth must be confirmed by a test. If measured the main bonding will need to be disconnected for the duration of the test.

* **Short-circuit capacity** of the device is noted, see Table 7.2A of the On-Site Guide or 2.7.15 of GN3

* **Prospective fault current (PFC).** The value recorded is the greater of either the short-circuit current or the earth fault current. Preferably determined by enquiry of the supplier.

The following tests, where relevant, shall be carried out in the following sequence:

Continuity of protective conductors, including main and supplementary bonding
Every protective conductor, including main and supplementary bonding conductors, should be tested to verify that it is continuous and correctly connected.

*6 **Continuity**
Where Test Method 1 is used, enter the measured resistance of the phase conductor plus the circuit protective conductor (R_1+ R_2).
See 10.3.1 of the On-Site Guide or 2.7.5 of GN3.
During the continuity testing (Test Method 1) the following polarity checks are to be carried out:
(a) every fuse and single-pole control and protective device is connected in the phase conductor only
(b) centre-contact bayonet and Edison screw lampholders have outer contact connected to the neutral conductor
(c) wiring is correctly connected to socket-outlets and similar accessories.
Compliance is to be indicated by a tick in polarity column 11.

*7 Where Test Method 2 is used, the maximum value of R_2 is recorded in column 7.
Where the alternative method of Regulation 413-02-12 is used for shock protection, the resistance of the circuit protective conductor R_2 is measured and recorded in column 7.
See 10.3.1 of the On-Site Guide or 2.7.5 of GN3.

*8 **Continuity of ring final circuit conductors**
A test shall be made to verify the continuity of each conductor including the protective conductor of every ring final circuit.
See 10.3.2 of the On-Site Guide or 2.7.6 of GN3.

*9, *10 **Insulation Resistance**
All voltage sensitive devices to be disconnected or test between live conductors (phase and neutral) connected together and earth.
The insulation resistance between live conductors is to be inserted in column 9.
The minimum insulation resistance values are given in Table 10.1 of the On-Site Guide or Table 2.2 of GN3.
See 10.3.3(iv) of the On-Site Guide or 2.7.7 of GN3.

All the preceding tests should be carried out before the installation is energised.

*11 **Polarity**
A satisfactory polarity test may be indicated by a tick in column 11.
Only in a Schedule of Test Results associated with a Periodic Inspection Report is it acceptable to record incorrect polarity.

*12 **Earth fault loop impedance Z_s**
This may be determined either by direct measurement at the furthest point of a live circuit or by adding (R_1 + R_2) of column 6 to Z_e. Z_e is determined by measurement at the origin of the installation or preferably the value declared by the supply company used.
$Z_s = Z_e + (R_1 + R_2)$. Z_s should be less than the values given in Appendix 2 of the On-Site Guide or App 2 of GN3.

*13 **Functional testing**
The operation of RCDs (including RCBOs) shall be tested by simulating a fault condition, independent of any test facility in the device.
Record operating time in column 13. Effectiveness of the test button must be confirmed.
See Section 11 of the On-Site Guide or 2.7.16 of GN3

*14 All switchgear and controlgear assemblies, drives, control and interlocks, etc must be operated to ensure that they are properly mounted, adjusted, and installed.
Satisfactory operation is indicated by a tick in column 14.

Earth electrode resistance
The earth electrode resistance of TT installations must be measured, and normally an RCD is required.
For reliability in service the resistance of any earth electrode should be below 200 Ω. Record the value on Form 1, 2 or 6, as appropriate. See 10.3.5 of the On-Site Guide or 2.7.13 of GN3.

NOTES ON COMPLETION OF MINOR ELECTRICAL INSTALLATION WORKS CERTIFICATE

Scope

The Minor Electrical Installation Works Certificate form is only to be used for additions to an electrical installation that do not extend to the introduction of a new circuit e.g. the addition of a socket-outlet or a lighting point to an existing circuit (Regulation 741-01-03).

Part 1 Description of minor works

1,2 The minor works must be so described that the work that is the subject of the certification can be readily identified.

4 See Regulations 120-01-03 and 120-02. No departures are to be expected except in most unusual circumstances. See also Regulation 743-01-01.

Part 2 Installation details

2 The method of protection against indirect contact shock must be clearly identified e.g. earthed equipotential bonding and automatic disconnection of supply using fuse/circuit-breaker/RCD.

4 If the existing installation lacks either an effective means of earthing or adequate main equipotential bonding conductors, this must be clearly stated. See Regulation 743-01-02.

Recorded departures from BS 7671 may constitute non-compliance with the Electricity Supply Regulations 1988 as amended or the Electricity at Work Regulations 1989. It is important that the client is advised immediately in writing.

Part 3 Essential Tests

The relevant provisions of Part 7 (Inspection and Testing) of BS 7671 must be applied in full to all minor works. For example, where a socket-outlet is added to an existing circuit it is necessary to:

1 establish that the earthing contact of the socket-outlet is connected to the main earthing terminal

2 measure the insulation resistance of the circuit that has been added to, and establish that it complies with Table 71A of BS 7671

3 measure the earth fault loop impedance to establish that the maximum permitted disconnection time is not exceeded

4 check that the polarity of the socket-outlet is correct

5 (if the work is protected by an RCD) verify the effectiveness of the RCD.

Part 4 Declaration

1,3 The Certificate shall be made out and signed by a competent person in respect of the design, construction, inspection and testing of the work.

1,3 The competent person will have a sound knowledge and experience relevant to the nature of the work undertaken and to the technical standards set down in BS 7671, be fully versed in the inspection and testing procedures contained in the Regulations and employ adequate testing equipment.

2 When making out and signing a form on behalf of a company or other business entity, individuals shall state for whom they are acting.

Form 5

MINOR ELECTRICAL INSTALLATION WORKS CERTIFICATE

(REQUIREMENTS FOR ELECTRICAL INSTALLATIONS - BS 7671 [IEE WIRING REGULATIONS])

To be used only for minor electrical work which does not include the provision of a new circuit

PART 1 : Description of minor works

1. Description of the minor works :..

2. Location/Address : ....

3. Date minor works completed :..

4. Details of departures, if any, from BS 7671

....

....

....

PART 2 : Installation details

1. System earthing arrangement: TN-C-S ☐ TN-S ☐ TT ☐

2. Method of protection against indirect contact: ....

3. Protective device for the modified circuit : Type BS .... Rating .... A

4. Comments on existing installation, including adequacy of earthing and bonding arrangements : (see Regulation 130-07)

....

....

....

PART 3 : Essential Tests

1. Earth continuity : satisfactory ☐

2. Insulation resistance:

   Phase/neutral .... MΩ

   Phase/earth .... MΩ

   Neutral/earth .... MΩ

3. Earth fault loop impedance .... Ω

4. Polarity : satisfactory ☐

5. RCD operation (if applicable) : Rated residual operating current $I_{\Delta n}$ ....mA and operating time of ....ms (at $I_{\Delta n}$)

PART 4 : Declaration

1. I/We CERTIFY that the said works do not impair the safety of the existing installation, that the said works have been designed, constructed, inspected and tested in accordance with BS 7671 : .... (IEE Wiring Regulations), amended to .... and that the said works, to the best of my/our knowledge and belief, at the time of my/our inspection, complied with BS 7671 except as detailed in Part 1.

2. Name: ....

   For and on behalf of: ....

   Address: ....

   ....

   ....

3. Signature: ....

   Position: ....

   Date: ....

MINOR ELECTRICAL INSTALLATION WORKS CERTIFICATE
GUIDANCE FOR RECIPIENTS (to be appended to the Certificate)

This Certificate has been issued to confirm that the electrical installation work to which it relates has been designed, constructed and inspected and tested in accordance with British Standard 7671, (The IEE Wiring Regulations.)

You should have received an original Certificate and the contractor should have retained a duplicate. If you were the person ordering the work, but not the owner of the installation, you should pass this Certificate, or a copy of it, to the owner.

The Minor Works Certificate is only to be used for additions, alterations or replacements to an installation that do not extend to the provision of a new circuit. Examples include the addition of a socket-outlet or lighting point to an existing circuit, or the replacement or relocation of a light switch. A separate Certificate should have been received for each existing circuit on which minor works have been carried out. This Certificate is not valid if you requested the contractor to undertake more extensive installation work. An Electrical Installation Certificate would be required in such circumstances.

The "original" Certificate should be retained in a safe place and be shown to any person inspecting or undertaking further work on the electrical installation in the future. If you later vacate the property, this Certificate will demonstrate to the new owner that the minor electrical installation work carried out complied with the requirements of British Standard 7671 at the time the Certificate was issued.

PERIODIC INSPECTION REPORT

NOTES:

1. This Periodic Inspection Report form shall only be used for the reporting on the condition of an existing installation.
2. The Report, normally comprising at least four pages, shall include schedules of both the inspection and the test results. Additional sheets of test results may be necessary for other than a simple installation. The page numbers of each sheet shall be indicated, together with the total number of sheets involved. The Report is only valid if a Schedule of Inspections and a Schedule of Test Results are appended.
3. The intended purpose of the Periodic Inspection Report shall be identified, together with the recipient's details in the appropriate boxes.
4. The maximum prospective fault current recorded should be the greater of either the short-circuit current or the earth fault current.
5. The 'Extent and Limitations' box shall fully identify the elements of the installation that are covered by the report and those that are not, this aspect having been agreed with the client and other interested parties before the inspection and testing is carried out.
6. The recommendation(s), if any, shall be categorised using the numbered coding 1-4 as appropriate.
7. The 'Summary of the Inspection' box shall clearly identify the condition of the installation in terms of safety.
8. Where the periodic inspection and testing has resulted in a satisfactory overall assessment, the time interval for the next periodic inspection and testing shall be given. The IEE Guidance Note 3 provides guidance on the maximum interval between inspections for various types of buildings. If the inspection and testing reveal that parts of the installation require urgent attention, it would be appropriate to state an earlier re-inspection date having due regard to the degree of urgency and extent of the necessary remedial work.
9. If the space available on the model form for information on recommendations is insufficient, additional pages shall be provided as necessary.

PERIODIC INSPECTION REPORT FOR AN ELECTRICAL INSTALLATION (note 1)

(REQUIREMENTS FOR ELECTRICAL INSTALLATIONS - BS 7671 [IEE WIRING REGULATIONS])

DETAILS OF THE CLIENT

Client: ....

Address: ....

Purpose for which this Report is required: .... (note 3)

DETAILS OF THE INSTALLATION Tick boxes as appropriate

Occupier: ....

Installation: ....

Address: ....

Description of Premises: Domestic ☐ Commercial ☐ Industrial ☐ Other ☐

....

Estimated age of the Electrical Installation: .... years

Evidence of Alterations or Additions: Yes ☐ No ☐ Not apparent ☐

If "Yes", estimate age: .... years

Date of last inspection: .... Records available Yes ☐ No ☐

EXTENT AND LIMITATIONS OF THE INSPECTION (note 5)

Extent of electrical installation covered by this report: ....

....

....

Limitations: ....

....

....

This inspection has been carried out in accordance with BS 7671 : 2001 (IEE Wiring Regulations), amended to .... Cables concealed within trunking and conduits, or cables and conduits concealed under floors, in roof spaces and generally within the fabric of the building or underground have not been inspected.

NEXT INSPECTION (note 8)

I/We recommend that this installation is further inspected and tested after an interval of not more than .... months/years, provided that any observations 'requiring urgent attention' are attended to without delay.

DECLARATION

INSPECTED AND TESTED BY

Name: .... Signature: ....

For and on behalf of: .... Position: ....

Address: ....

.... Date: ....

....

SUPPLY CHARACTERISTICS AND EARTHING ARRANGEMENTS Tick boxes and enter details, as appropriate

Earthing arrangements

TN-C ☐
TN-S ☐
TN-C-S ☐
TT ☐
IT ☐

Alternative source ☐ of supply (to be detailed on attached schedules)

Number and Type of Live Conductors

a.c. ☐ d.c. ☐
1-phase, 2-wire ☐ 2-pole ☐
1-phase, 3 wire ☐ 3-pole ☐
2-phase, 3-wire ☐ other ☐
3-phase, 3-wire ☐
3-phase, 4-wire ☐

Nature of Supply Parameters

Nominal voltage, U/U_o (1) V
Nominal frequency, f (1) .... Hz
Prospective fault current, I_{pf} (2) .... kA (note 4)
External loop impedance, Z_e (2) .... Ω

(Note: (1) by enquiry, (2) by enquiry or by measurement)

Supply Protective Device Characteristics

Type: ....
....

Nominal current rating ....A

PARTICULARS OF INSTALLATION REFERRED TO IN THE REPORT Tick boxes and enter details, as appropriate

Means of Earthing

Supplier's facility ☐
Installation earth electrode ☐

Details of Installation Earth Electrode (*where applicable)*

Type (e.g. rod(s), tape etc)	Location	Electrode resistance to earth
....	....	.... Ω

Main Protective Conductors

Earthing conductor: material .... csa ....mm^2 connection verified ☐
Main equipotential bonding conductors material .... csa ....mm^2 connection verified ☐

To incoming water service ☐ To incoming gas service ☐ To incoming oil service ☐ To structural steel ☐
To lightning protection ☐ To other incoming service(s) ☐ (state details...)

Main Switch or Circuit-breaker

BS, Type....................................... No. of poles .... Current rating ....A Voltage rating V

Location... Fuse rating or setting ....A

Rated residual operating current $I_{\Delta n}$ = ... mA, and operating time of .... ms (at $I_{\Delta n}$) (applicable only where an RCD is suitable and is used as a main circuit-breaker)

OBSERVATIONS AND RECOMMENDATIONS Tick boxes as appropriate (note 9)

Recommendations as detailed below note 6

Referring to the attached Schedule(s) of Inspection and Test Results, and subject to the limitations specified at the Extent and Limitations of the Inspection section

☐ No remedial work is required ☐ The following observations are made:

....
....
....
....
....
....
....

One of the following numbers, as appropriate, is to be allocated to each of the observations made above to indicate to the person(s) responsible for the installation the action recommended.

1 requires urgent attention 2 requires improvement 3 requires further investigation

4 does not comply with BS 7671: 2001 amended to .... This does not imply that the electrical installation inspected is unsafe.

SUMMARY OF THE INSPECTION (note 7)

Date(s) of the inspection: ....

General condition of the installation: ....

....
....
....

Overall assessment: Satisfactory/Unsatisfactory (note 8)

SCHEDULE(S)

The attached Schedules are part of this document and this Report is valid only when they are attached to it.

.... Inspection Schedules and .... Test Result Schedules are attached.

(Enter quantities of schedules attached).

Page 2 of ....

PERIODIC INSPECTION REPORT
GUIDANCE FOR RECIPIENTS (to be appended to the Report)

This Periodic Inspection Report form is intended for reporting on the condition of an existing electrical installation.

You should have received an original Report and the contractor should have retained a duplicate. If you were the person ordering this Report, but not the owner of the installation, you should pass this Report, or a copy of it, immediately to the owner.

The original Report is to be retained in a safe place and be shown to any person inspecting or undertaking work on the electrical installation in the future. If you later vacate the property, this Report will provide the new owner with details of the condition of the electrical installation at the time the Report was issued.

The 'Extent and Limitations' box should fully identify the extent of the installation covered by this Report and any limitations on the inspection and tests. The contractor should have agreed these aspects with you and with any other interested parties (Licensing Authority, Insurance Company, Building Society etc) before the inspection was carried out.

The Report will usually contain a list of recommended actions necessary to bring the installation up to the current standard. **For items classified as 'requires urgent attention', the safety of those using the installation may be at risk,** and it is recommended that a competent person undertakes the necessary remedial work without delay.

For safety reasons, the electrical installation will need to be re-inspected at appropriate intervals by a competent person. The maximum time interval recommended before the next inspection is stated in the Report under 'Next Inspection.'

The Report is only valid if a Schedule of Inspections and a Schedule of Test Results are appended.

APPENDIX 1 — RESISTANCE OF COPPER AND ALUMINIUM CONDUCTORS

(This appendix is similar to Appendix 9 of the On-Site Guide).

To check compliance with Regulation 434-03-03 and/or Regulation 543-01-03, i.e. to evaluate the equation $S^2 = I^2 t/k^2$, it is necessary to establish the impedances of the circuit conductors to determine the fault current I and hence the protective device disconnection time t. 434-03-03 543-01-03

Fault current $I = U_o/Z_s$

where

U_o is the nominal voltage to earth,

Z_s is the earth fault loop impedance.

$Z_s = Z_e + R_1 + R_2$

where

Z_e is that part of the earth fault loop impedance external to the circuit concerned,

R_1 is the resistance of the phase conductor from the origin of the circuit to the point of utilization,

R_2 is the resistance of the protective conductor from the origin of the circuit to the point of utilization.

Similarly, in order to design circuits for compliance with BS 7671 and the limiting values of earth fault loop impedance given in Tables 41B1, 41B2 and 41D of BS 7671, or for compliance with the limiting values of the circuit protective conductor given in Table 41C, it is necessary to establish the relevant impedances of the circuit conductors concerned at their operating temperature.

Table 1A gives values of (R1 + R2) per metre for various combinations of conductors up to and including 50 mm^2 cross-sectional area. It also gives values of resistance (milliohms) per metre for each size of conductor. These values are at 20 °C.

TABLE 1A
Values of resistance/metre for copper and aluminium conductors and of R1 + R2 per metre at 20 °C in milliohms/metre

Cross-sectional area (mm²)		Resistance/metre or (R1 + R2)/metre (mΩ/m)	
Phase conductor	Protective conductor	Copper	Aluminium
1	—	18.10	
1	1	36.20	
1.5	—	12.10	
1.5	1	30.20	
1.5	1.5	24.20	
2.5	—	7.41	
2.5	1	25.51	
2.5	1.5	19.51	
2.5	2.5	14.82	
4	—	4.61	
4	1.5	16.71	
4	2.5	12.02	
4	4	9.22	
6	—	3.08	
6	2.5	10.49	
6	4	7.69	
6	6	6.16	
10	—	1.83	
10	4	6.44	
10	6	4.91	
10	10	3.66	
16	—	1.15	1.91
16	6	4.23	—
16	10	2.98	—
16	16	2.30	3.82
25	—	0.727	1.20
25	10	2.557	—
25	16	1.877	—
25	25	1.454	2.40
35	—	0.524	0.87
35	16	1.674	2.78
35	25	1.251	2.07
35	35	1.048	1.74
50	—	0.387	0.64
50	25	1.114	1.84
50	35	0.911	1.51
50	50	0.774	1.28

TABLE 1B
Ambient temperature multipliers (α) to Table 1A

Expected ambient temperature	Correction factor (see note)
0 °C	0.92
5 °C	0.94
10 °C	0.96
15 °C	0.98
20 °C	1.00
30 °C	1.04
40 °C	1.08

Note:
The correction factor is given by: {1 + 0.004 (ambient temp - 20 °C)} where 0.004 is the simplified resistance coefficient per °C at 20 °C given by BS 6360 for copper and aluminium conductors.

For verification purposes the designer will need to give the values of the phase and circuit protective conductor resistances at the ambient temperature expected during the tests. This may be different from the reference temperature of 20 °C used for Table 1A. The correction factors in Table 1B may be applied to the Table 1A values to take account of the ambient temperature (for test purposes only).

Standard overcurrent devices

Table 1C gives the multipliers to be applied to the values given in Table 1A for the purpose of calculating the resistance at maximum operating temperature of the phase conductors and/or circuit protective conductors in order to determine compliance with, as applicable:

(a) earth fault loop impedance of Tables 41B1, 41B2 or 41D of BS 7671 Table 41B1 Table 41B2 Table 41D

(b) earth fault loop impedance and resistance of protective conductor of Table 41C of BS 7671. Table 41C

Where it is known that the actual operating temperature under normal load conditions is less than the maximum permissible value for the type of cable insulation concerned (as given in the Tables of current-carrying capacity) the multipliers given in Table 1C may be reduced accordingly.

TABLE 1C Conductor temperature factor F for standard devices

Table 41B1
Table 41B2
Table 41C
Table 41D

Multipliers to be applied to Table 1A for devices in Tables 41B1, 41B2, 41C and 41D

Conductor Installation	Conductor Insulation		
	70 °C thermoplastic (pvc)	85 °C thermosetting note 4	90 °C thermosetting note 4
Not incorporated in a cable and not bunched - notes 1, 3	1.04	1.04	1.04
Incorporated in a cable or bunches - notes 2, 3	1.20	1.26	1.28

Table 54B

Table 54C

Note 1 See Table 54B of BS 7671. These factors apply when protective conductor is not incorporated or bunched with cables, or for bare protective conductors in contact with cable covering.

Table 54B

Note 2 See Table 54C of BS 7671. These factors apply when the protective conductor is a core in a cable or is bunched with cables.

Table 54C

Note 3 The factors are given by
F = 1 + 0.004{conductor operating temperature - 20 °C}
where 0.004 is the simplified resistance coefficient per °C at 20 °C given in BS 6360 for copper and aluminium conductors.

Note 4 If cable loading is such that the maximum operating temperature is 70 °C, thermoplastic (70 °C) factors are appropriate.

APPENDIX 2 — MAXIMUM PERMISSIBLE MEASURED EARTH FAULT LOOP IMPEDANCE

(The tables in this appendix are as Appendix 2 of the On-Site Guide.)

The Tables provide maximum permissible measured earth fault loop 713-11
impedances (Zs) for compliance with BS 7671 where conventional final 413-02-05
circuits are used. The values are those that must not be exceeded in 413-02-10
the tests described in Section 2.7.14 at a test ambient temperature of 413-02-11
10 °C to 20 °C. Table 2E provides correction factors for other ambient 413-02-14
temperatures. 543-01-03

When the cables to be used are to Table 5 and Table 6 of BS6004 or table 7 of BS7211, or are other thermoplastic (pvc) or lsf cables to these British Standards and if the cable loading is such that the maximum operating temperature is 70 °C, then Tables 2A, 2B and 2C give the maximum earth fault loop impedances for circuits with:

(a) protective conductors of copper and having from 1 mm² to 16 mm² cross-sectional area,

(b) where the overcurrent protective device is a fuse to BS 88-2.1, BS 88-6, BS 1361 or BS 3036.

For each type of fuse, two tables are given:

- where the circuit concerned feeds socket-outlets and the disconnection time for compliance with Regulation 413-02-09 is 0.4 s, and 413-02-09

- where the circuit concerned feeds fixed equipment only and the disconnection time for compliance with Regulation 413-02-13 is 5 s. 413-02-13

In each table the earth fault loop impedances given correspond to the appropriate disconnection time from a comparison of the time/current characteristic of the device concerned and the equation given in Regulation 543-01-03. 543-01-03

The tabulated values apply only when the nominal voltage to Earth (U_o) is 230 V.

Table 2D gives the maximum measured Z_s for circuits protected by circuit-breakers to BS EN 60898, RCBOs to BS EN 61009 and circuit-breakers to BS 3871.

Note: The impedances tabulated in this Appendix are maximum measured values at an assumed conductor temperature of 10 ° C. These values are lower than those in the tables in BS 7671, which are design figures at the conductor normal operating temperature.

TABLE 2A
Maximum measured earth fault loop impedance (in ohms) when overcurrent protective device is a semi-enclosed fuse to BS 3036

(see Note)

(i) 0.4 second disconnection

413-02-05
Table 41B1
543-01-03

Protective conductor (mm^2)	Fuse rating (amperes)				
	5	15	20	30	45
1.0	8.00	2.14	1.48	NP	NP
1.5	8.00	2.14	1.48	0.91	NP
2.5 to 16.0	8.00	2.14	1.48	0.91	0.50

(ii) 5 seconds disconnection

413-02-05
Table 41D
543-01-03

Protective conductor (mm^2)	Fuse rating (amperes)				
	5	15	20	30	45
1.0	14.80	4.46	2.79	NP	NP
1.5	14.80	4.46	3.20	2.08	NP
2.5	14.80	4.46	3.20	2.21	1.20
4.0 to 16.0	14.80	4.46	3.20	2.21	1.33

Note: A value of k of 115 from Table 54C of BS 7671 is used. This is suitable for pvc insulated and sheathed cables to Table 5 or Table 6 of BS 6004 and for lsf insulated and sheathed cables to Table 7 of BS 7211. The k value is based on both the thermoplastic (pvc) and thermosetting (lsf) cables operating at a maximum operating temperature of 70 °C. Table 54C

NP protective conductor, fuse combination NOT PERMITTED.

TABLE 2B
Maximum measured earth fault loop impedance (in ohms) when overcurrent protective device is a fuse to BS 88-2.1 or BS 88-6

(see Note)

(i) 0.4 second disconnection

413-02-05
Table 41B1
543-01-03

Protective conductor (mm^2)	Fuse rating (amperes)							
	6	10	16	20	25	32	40	50
1.0	7.11	4.26	2.26	1.48	1.20	0.69	NP	NP
1.5	7.11	4.26	2.26	1.48	1.20	0.87	0.67	NP
2.5 to 16.0	7.11	4.26	2.26	1.48	1.20	0.87	0.69	0.51

(ii) 5 seconds disconnection

413-02-05
Table 41D
543-01-03

Protective conductor (mm^2)	Fuse rating (amperes)							
	6	10	16	20	25	32	40	50
1.0	11.28	6.19	3.20	1.75	1.24	0.69	NP	NP
1.5	11.28	6.19	3.49	2.43	1.60	1.12	0.67	NP
2.5	11.28	6.19	3.49	2.43	1.92	1.52	1.13	0.56
4.0	11.28	6.19	3.49	2.43	1.92	1.52	1.13	0.81
6.0 to 16.0	11.28	6.19	3.49	2.43	1.92	1.52	1.13	0.87

Note: A value of k of 115 from Table 54C of BS 7671 is used. This is suitable for pvc insulated and sheathed cables to Table 5 or Table 6 of BS 6004 and for lsf insulated and sheathed cables to Table 7 of BS 7211. The k value is based on both the thermoplastic (pvc) and thermosetting (lsf) cables operating at a maximum operating temperature of 70 °C. Table 54C

NP protective conductor, fuse combination NOT PERMITTED.

TABLE 2C
Maximum measured earth fault loop impedance (in ohms) when overcurrent protective device is a fuse to BS 1361 (see Note)

(i) 0.4 second disconnection

413-02-05
Table 41B1
543-01-03

Protective conductor (mm^2)	Fuse rating (amperes)				
	5	15	20	30	45
1.0	8.72	2.74	1.42	0.80	NP
1.5	8.72	2.74	1.42	0.96	0.34
2.5 to 16.0	8.72	2.74	1.42	0.96	0.48

(ii) 5 seconds disconnection

413-02-05
Table 41D
543-01-03

Protective conductor (mm^2)	Fuse rating (amperes)				
	5	15	20	30	45
1.0	13.68	4.18	1.75	0.80	NP
1.5	13.68	4.18	2.24	1.20	0.34
2.5	13.68	4.18	2.34	1.54	0.53
4.0	13.68	4.18	2.34	1.54	0.70
6.0 to 16.0	13.68	4.18	2.34	1.54	0.80

Note: A value of k of 115 from Table 54C of BS 7671 is used. This is suitable for pvc insulated and sheathed cables to Table 5 or Table 6 of BS 6004 and for lsf insulated and sheathed cables to Table 7 of BS 7211. The k value is based on both the thermoplastic (pvc) and thermosetting (lsf) cables operating at a maximum operating temperature of 70 °C. Table 54C

NP protective conductor, fuse combination NOT PERMITTED.

TABLE 2D

Maximum measured earth fault loop impedance (in ohms) when overcurrent protective device is a circuit-breaker to BS EN 60898 or an RCBO to BS EN 61009

(i) both 0.4 and 5 seconds disconnection times (see Notes 1 and 2)

circuit-breaker type	circuit-breaker rating (amperes)											
	6	10	16	20	25	32	40	50	63	80	100	125
B	6.40	3.84	2.40	1.92	1.54	1.20	0.96	0.77	0.61	0.48	0.38	0.30
C	3.20	1.92	1.20	0.96	0.77	0.60	0.48	0.38	0.30	0.24	0.19	0.15
D	1.60	0.96	0.60	0.48	0.38	0.30	0.24	0.19	0.15	0.12	0.09	0.08

Maximum measured earth fault loop impedance (in ohms) when overcurrent protective devices is an mcb to BS 3871

(ii) both 0.4 and 5 seconds disconnection times (see Note 1)

mcb type	mcb rating (amperes)												
	5	6	10	15	16	20	25	30	32	40	45	50	63
1	9.60	8.00	4.80	3.20	3.00	2.40	1.92	1.60	1.50	1.20	1.06	0.96	0.76
2	5.48	4.57	2.74	1.83	1.71	1.37	1.09	0.92	0.86	0.69	0.61	0.55	0.43
3	3.84	3.20	1.92	1.28	1.20	0.96	0.77	0.64	0.60	0.48	0.42	0.38	0.30

Notes:

1) A value of k of 115 from Table 54C of BS 7671 is used. This is suitable for pvc insulated and sheathed cables to Table 5 or Table 6 of BS 6004 and for lsf insulated and sheathed cables to Table 7 of BS 7211. The k value is based on both the thermoplastic (pvc) and thermosetting (lsf) cables operating at a maximum operating temperature of 70 °C.

2) In TN systems it is preferable for reliable operation for indirect shock protection to be provided by overcurrent devices including RCBOs operating as overcurrent devices; that is, with loop impedance complying with the table above. RCBOs are then providing indirect shock protection as circuit-breakers (voltage independent devices), and supplementary protection against direct contact (voltage dependent) as RCDs. 413-02-04 471-16

TABLE 2E
Ambient temperature correction factors

Ambient temperature °C	Correction factors (from 10 °C) notes 1, 2
0	0.96
5	0.98
10	1.00
15	1.02
20	1.04
25	1.06
30	1.08

Notes:

1 - The correction factor is given by: {1 + 0.004 (Ambient temp - 10)} where 0.004 is the simplified resistance coefficient per °C at 20 °C given by BS 6360 for both copper and aluminium conductors

2 - The factors are different to those of Table 1B because Table 2E corrects from 10 °C and Table 1B from 20 °C. The values in Tables 2A to 2D are for a 10 °C ambient.

The appropriate ambient correction factor from Table 2E is applied to the earth fault loop impedances of Tables 2A to 2D if the ambient temperature is other than 10 °C. For example, if the ambient temperature is 25 °C the measured earth fault loop impedance of a circuit protected by a 32 A type B circuit-breaker to BS EN 60898 should not exceed 1.20 x 1.06 = 1.27 Ω.

Methods of adjusting tabulated values of Z_S

(see Section 2.7.14, items 2 and 3)

A circuit is wired in flat twin and cpc 70 °C thermoplastic (pvc) cable and protected by a 6 amp type B circuit-breaker to BS EN 60898. When tested at an ambient temperature of 20 °C, determine the maximum acceptable measured value of Z_S for the circuit.

Solution:

$$Z_{test\ (max)} = \frac{1}{F} Z_S$$

From Table 41B2(e) of BS 7671, the maximum permitted value of Z_S = 8 ohms

From Table 1C in Appendix 1 of this Guidance Note, F = 1.20

$$Z_{test\ (max)} = \frac{1}{1.20} \times 8$$

$$Z_{test\ (max)} = 6.67 \text{ ohms}$$

A more accurate value can be obtained if the external earth fault loop impedance, Z_e, is known. In this case, the following formula may be used:

$$Z_{test} \leq \frac{1}{F}\{Z_S + Z_e(F - 1)\}$$

In the example above, assume Z_e is 0.35. Thus, the more accurate value is:

$$Z_{test\ (max)} = \frac{1}{1.20}\{8 + 0.35(1.20 - 1)\}$$

$$Z_{test\ (max)} = 6.72 \text{ ohms}$$

Where the test ambient temperature is likely to be other than 20 °C, a further correction can be made to convert the value to the expected ambient temperature, using the following formula:

$$Z_{test\ (max)} = \frac{1}{F + 1 - \alpha}\{Z_S + Z_e(F - \alpha)\}$$

where: α is given by Table 1B of Appendix 1.

In the example above, assume the ambient temperature is 5 °C.

From Table 1B in Appendix 1 of this Guidance Note, $\alpha = 0.94$

Thus, the accurate reading including temperature compensation is:

$$Z_{test\ (max)} = \frac{1}{1.20 + 1 - 0.94}\{8 + 0.35(1.20 - 0.94)\}$$

$Z_{test\ (max)} = 0.79\ \{8 + 0.091\}$

$Z_{test\ (max)} = 6.42$ ohms

Alternatively, a conductor temperature resistance factor, F, of 1.26, which corresponds to 5 °C, can be used instead of the 1.20 factor in the formula shown in the second solution box. This gives the same result as that shown above.

Note

If reduced cross-sectional area protective conductors are used, maximum earth fault loop impedances may need to be further reduced to ensure disconnection times are sufficiently short to prevent overheating of protective conductors during earth faults. The requirement of the equation in Regulation 543-01-03 needs to be met :

$$S \geq \frac{\sqrt{I^2 t}}{k}$$

where:

	S	is the nominal cross-sectional area of the conductor in mm^2
	≥	means greater than or equal to
	k	is a factor from Tables 54B, C or D
	I	is the prospective earth fault current given by U_{oc}/Z_s
	Z_s	is the loop impedance at conductor normal operating temperature
	t	is the operating time of the overcurrent device in seconds. This is obtained from the graphs in Appendix 3 of BS 7671, as the prospective earth fault current I (= U_{oc}/Z_s) is known.

The following example illustrates how measurements taken at 20 °C may be adjusted to 70 °C values, taking the $R_1 + R_2$ reading for the circuit into account.

In the previous example, taking the $(R_1 + R_2)$ reading for the circuit as 0.2 ohm:

Z_s for the circuit at 70 °C

$= Z_e + F(R_1 + R_2)_{test}$

$= 0.35 + 1.20\ (0.2)$

$= 0.59$ ohm

The temperature corrected Z_s figure of 0.59 ohm is acceptable, since it is less than the maximum value of 8 ohms given in Table 41B2 of BS 7671.

The formula above involves taking measurements at 20 °C and converting them to 70 °C values. Alternatively, the 70 °C values can be converted to the values at the expected ambient temperature, e.g. 20 °C, when the measurement is carried out.

Taking the same circuit,

$Z_{test} = Z_e + (R_1 + R_2)_{test}$

$= 0.35 + 0.2$

$= 0.55$ ohm

From the formula $Z_{test\ (max)} = \frac{1}{F} Z_s$

$Z_{s\ (max)}$ from BS 7671 = 8 ohms

$Z_{test\ (max)} = \frac{1}{1.20} \times 8 = 6.67$ ohms

Therefore, as 0.55 is less than 6.67, the circuit is acceptable.

Index

Guidance Note 3
Inspection & Testing
4th edition 2002, ISBN 0 85296 991 0

This ERRATA have been corrected in this reprint.
This page is provided to help trainers and lecturers.

ERRATA (March 2002)

Table 2D(ii), Page 114, Delete last column headed I_n as follows:

45	50	63	~~I_n~~
1.06	0.96	0.76	~~$60/I_n$~~
0.61	0.55	0.43	~~$34.3/I_n$~~
0.42	0.38	0.30	~~$24/I_n$~~

Errata published by IEE, 9 March 2002

NOTES

NOTES

NOTES

NOTES